EDITED BY F. ROBERT VAN DER LINDEN

BEST OF THE NATIONAL AIR AND SPACE MUSEUM

✸ Smithsonian Books

An Imprint of HarperCollinsPublishers

i: Overhead view of the nose of the Boeing 307 as the aircraft is moved into the Udvar-Hazy Center. **ii:** The Wright 1903 Flyer in the Milestones of Flight gallery. **iv:** Museum specialists push the Boeing 307 into the Udvar-Hazy Center. **vi:** Space Race gallery. **viii-ix:** *(left to right)* Bell X-1, Blériot XI, Republic P-47D Thunderbolt. **x-xi:** *(left to right)* Boeing P-26A, Boeing B-29 Superfortress *Enola Gay*, Gemini 4. **xii:** The Boeing Aviation Hangar at the National Air and Space Museum's Steven F. Udvar-Hazy Center. **xiv:** The Concorde supersonic transport.

BEST OF THE NATIONAL AIR AND SPACE MUSEUM. Copyright © 2006 by the Smithsonian Institution. All rights reserved. Printed in the United States of America. No part of this book may be used or reproduced in any manner whatsoever without written permission except in the case of brief quotations embodied in critical articles and reviews. For information address HarperCollins Publishers, 10 East 53rd Street, New York, NY 10022.

HarperCollins books may be purchased for educational, business, or sales promotional use. For information please write: Special Markets Department, HarperCollins Publishers Inc., 10 East 53rd Street, New York, NY 10022.

Book design by Judith Stagnitto Abbate / Abbate Design

First Smithsonian Books edition published 2006.
Library of Congress Cataloging-in-Publication data has been applied for.

ISBN-10: 0-06-085155-4
ISBN-13: 978-0-06-085155-2

02 03 04 05 06 ❖/QW 10 9 8 7 6 5 4 3 2 1

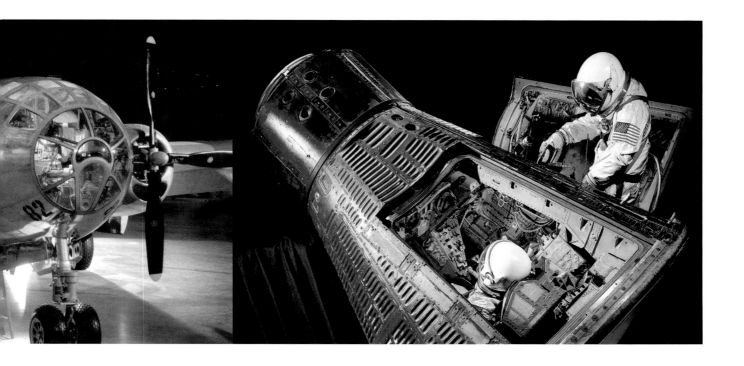

THE SMITHSONIAN Institution's National Air and Space Museum is fortunate to preserve the finest and most complete collection of aviation and space artifacts in the world. Since 1882, when the first objects were collected—a set of Chinese kites from the Centennial Exhibition—the Smithsonian has endeavored to preserve the history of flight as well as participate directly in it. Indeed the pioneering work of third secretary Samuel Pierpont Langley was a milestone on the road to the creation of the first practical aircraft. The Smithsonian's direct participation in the creation of the National Advisory Committee for Aeronautics (NACA)—the predecessor to NASA—and its support of the early work of rocket pioneer Robert Goddard was instrumental in promoting the technical development of aviation and spaceflight.

Beginning in 1920, a man of diminutive appearance but a giant in stature began his work for the Smithsonian. For the next five decades, Paul Edward Garber vigorously sought out and acquired the most significant aircraft and other objects he could find. The results were astounding. Because of his untiring efforts, and those of the curators who followed him, the National Air and Space Museum houses the majority of the world's most significant aerospace artifacts.

In the pages that follow, you will see 100 of the premier objects currently on display from our collection. From the original 1903 Wright Flyer to the space shuttle and so much in between, the National Air and Space Museum preserves and exhibits these priceless treasures at our flagship building on the National Mall in our nation's capital and now also at our latest facility, the new Steven F. Udvar-Hazy Center at Washington Dulles International Airport in nearby Chantilly, Virginia. In this way the Smithsonian Institution is well able to increase and diffuse our knowledge of aviation and spaceflight to millions of visitors each year.

John R. Dailey
Director

THE SMITHSONIAN Institution's National Air and Space Museum (NASM) preserves over 35,000 aerospace artifacts—and that number grows steadily each year. Singling out 75 aviation and 25 space objects from this unparalleled collection of aerospace artifacts is an extremely difficult task and one that will undoubtedly remain open to understandable second-guessing. To describe in depth every significant artifact would result in a massive book too expensive to buy and too heavy to lift. Consequently, the contents of this book have been chosen carefully and in the full knowledge that some truly significant objects have been excluded. Those included are all currently on display either at the main NASM building on the National Mall or at our newest location, the Steven F. Udvar-Hazy Center at Washington Dulles International Airport. Those excluded include aircraft and spacecraft currently on loan to museums around the world, objects awaiting restoration, and artifacts planned for display but still in storage. That is why the Curtiss NC-4, and the Saturn V, among others, are not in this work.

All of the essays were written by current and former members of the NASM curatorial staff. Although edited to fit the format for this book, each essay represents the work of the staff member responsible for these objects. Currently from the Aeronautics Division, Division of Space History, and the Center for Earth and Planetary Studies, those individuals include: John Anderson, Paul Ceruzzi, Dorothy Cochrane, Martin Collins, Roger Conner, Robert Craddock, Tom Crouch, Dik Daso, R. E. G. Davies, David DeVorkin, Thomas Dietz, Von Hardesty, Peter Jakab, Jeremy Kinney, Roger Launius, Russell Lee, Cathy Lewis, Donald Lopez, Valerie Neal, Alan Needell, Michael Neufeld, Dominick Pisano, Alex Spencer, Robert van der Linden, Thomas Watters, Frank Winter, and Mandy Young.

Special thanks must be extended to Trish Graboske for her diligent work as NASM chief of publications in shepherding this and so many other manuscripts through the publication process. Melissa Keiser of the Archives Division traveled the extra mile to process the hundreds of photographs necessary for this book. Most important, were it not for the photograph research of Alex Spencer and his command of the intricacies of modern computers, this book would not have been completed.

THE M6A1 SEIRAN was designed as a bomber that could operate exclusively from a submarine and strike the United States directly. No Seiran ever saw combat, but they represent an ingenious blend of aviation and marine technology.

In 1942 the Japanese navy issued orders to build a new series of submarine aircraft carriers called the I-400 class. Navy planners envisioned a large fleet but only three were completed. An I-400 ship displaced 5,970 metric tons (6,560 tons) submerged, and was capable of traveling 60,000 km (43,000 mi.) while carrying three Seirans in waterproof compartments. The smaller AM-class Japanese submarines were also modified to carry two Seirans.

Soon after commencing the I-400 program, the navy directed Aichi to develop the prototype special attack aircraft M6A1. Aichi's chief engineer, Tashio Ozaki, confronted an ambitious challenge: develop an aircraft to haul a 250-kg (400-lb.) bomb, or an 800-kg (1,288-lb.) bomb or torpedo, and fly at least 474 km/h (294 mph) with jettisonable floats in place, or 559 km/h (347 mph) without floats. The navy also stipulated that assembling and launching the M6A1s should require no more than 30 minutes. To fit inside a 3.5-m (11-ft., 6-in.) diameter, cylinder-shaped hangar, Ozaki designed the main wing spar to rotate 90 degrees once the deck crew removed the two floats. After rotating the wings, the crew folded them back to lie flat against the fuselage. About two-thirds of each side of the horizontal stabilizer and the tip of the vertical stabilizer also folded down. Deck crews stored the floats and their support pylons in separate compartments.

Aichi completed the first prototype in October 1943. The navy was so pleased with the initial results that it ordered production to start even before Aichi delivered the remaining

DIMENSIONS: ————————

WINGSPAN 12.262 M (40 FT., 2.75 IN.)
LENGTH 11.63 M (38 FT., 2.25 IN.)
HEIGHT 4.58 M (15 FT.)
WEIGHT 3,301 KG (7,277 LB.)

1. The museum's Seiran on display at the Udvar-Hazy Center **2.** View looking directly forward of aft cockpit of the museum's restored Aichi M6A1 Seiran **3.** An Aichi M6A1 Seiran ("Clear Sky Storm") in a handling cradle. This Japanese navy special-attack bomber was designed to operate from a submarine.

prototypes. However, progress virtually stopped after a major earthquake severely disrupted the production line in December 1944. Boeing B-29 bomber raids further disrupted the project.

In March 1945, near the end of the war, the Japanese navy curtailed their submarine program. The first I-400 was finished on December 30, 1944, and the *I-401* followed a week later. With the submarine fleet now reduced, the navy required fewer Seirans so this program was also curtailed. Using parts on hand, Aichi eventually built twenty-six Seirans (including prototypes) and two Nanzan trainers.

Navy leaders organized the 1st Submarine Flotilla and 631st Air Corps and placed Capt. Tatsunoke Ariizumi in command of both units. The combined force consisted of the submarine carriers *I-400* and *I-401,* two AM-class submarines, the *I-13* and *I-14,* and ten Seiran bombers. The units practiced hard to reduce the assembly time for the Seiran, and the crews could launch three aircraft (without floats) in less than 15 minutes. The major drawback for flying without floats was that the pilot had to ditch the bomber near the submarine and await rescue. The aircraft was obviously lost.

Japanese navy planners chose to strike the locks of the Panama Canal using the 631st. The plan was for ten Seirans to strike the Gatun locks with six torpedoes and four bombs. Allied naval supremacy forced the Japanese to scuttle the attack plan. As an alternative, the Japanese decided to strike at the US Navy fleet anchored at Ulithi Atoll. On June 25, 1945, Ariizumi received orders for Operation Hikari. The six Seirans were to carry out kamikaze attacks on the American aircraft carriers and troop transports.

Trouble dogged the entire operation. The *I-13,* with two MYRTs aboard, was damaged by air attacks then sunk by a US destroyer. The *I-400* missed a crucial radio message from Ariizumi's flagship and proceeded to the wrong rendezvous point. On August 16, 1945, Ariizumi's flotilla received word that the war was over, and they were ordered to return to Japan. The Seirans, which had been disguised with American markings, were hastily repainted and then pushed into the sea.

The National Air and Space Museum's M6A1 was the last airframe built, and today it remains the only existing Seiran. It was transferred to NASM from the US Navy in 1962.

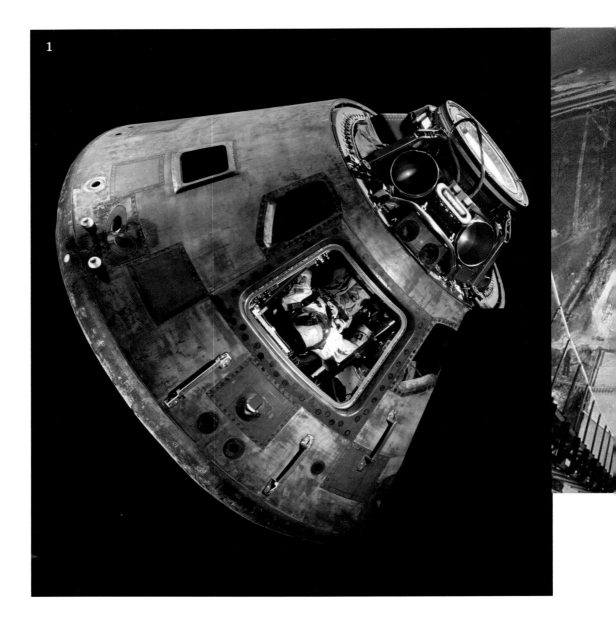

1

ON JULY 20, 1969, the crew of Apollo 11 fulfilled President John F. Kennedy's challenge to land a man on the Moon and safely return him to Earth. The culmination of the intense space race between the United States and the Soviet Union, this trip was a technological and political victory for the United States.

The Apollo 11 command module *Columbia* was the living quarters for the three-man crew during most of the first manned lunar landing mission. On July 16, Neil Armstrong, Edwin "Buzz" Aldrin, and Michael Collins climbed into *Columbia* for their eight-day journey. The command module was one of three parts of the Apollo 11 spacecraft. The other two were the service module and the lunar module.

The service module contained the main spacecraft propulsion system and consumables—oxygen, water, propellants, and hydrogen. The lunar module *Eagle* was the part Armstrong and Aldrin used to descend to the Moon's surface. The command module is the only portion of the spacecraft that returned to Earth.

For the launch, the lunar module was stored in a cone-shaped adapter between the service module and the Saturn V launch vehicle. Once the spacecraft was on its way to the

DIMENSIONS: ═══════

DIAMETER	3.2 M (10 FT., 2.25 IN.)
LENGTH	3.9 M (12 FT., 10 IN.)
WEIGHT	5,900 KG (13,000 LB.)

1. The Apollo 11 command module *Columbia* on exhibit in the National Air and Space Museum's Milestones of Flight gallery. **2.** 9:32 a.m. July 16, 1969, Apollo 11 lifts off from Kennedy Space Center Launch Complex 39A. **3.** A view of the Apollo 11 command module cockpit.

Moon, the command and service modules (CSM) pulled away from the adapter, turned around, and docked with the lunar lander. After the CSM/lunar lander combination reached the Moon, Armstrong and Aldrin entered the lunar module and undocked from *Columbia*. Collins remained in lunar orbit while his crewmates landed on the Moon's surface.

Following their historic landing and exploration of the Moon's surface, Armstrong and Aldrin rejoined Collins aboard *Columbia*. Collins, as command module pilot, fired the CSM's large engine and headed back to Earth. Several days later, on July 24, they discarded the service module and entered Earth's atmosphere.

The exterior of *Columbia* is covered with an epoxy-resin ablative heat shield, which prevented the module from burning and vaporizing when it entered the atmosphere at a speed of 40,000 km/h (25,000 mph) with an exterior temperature of 2,760°C (5,000°F). *Columbia* finished its flight with a parachute landing in the Pacific Ocean, where the USS *Hornet* retrieved it and its crew.

The cone-shaped spacecraft is divided into three compartments: forward, crew, and aft. The forward compartment is at the cone's apex, the crew compartment is in the center, and the aft compartment is in the base, or blunt end, of the craft. The forward compartment contained the parachutes and recovery equipment around the tunnel and hatch for passage to and from the lunar module. The crew compartment has a volume of 5.9 cu m (210 cu. ft.). It contains three couches for the crew during launch and landing. The couches are arranged so that each astronaut faces the main instrument panel. During flight, the astronauts folded the couches up to make more room in the spacecraft. Near the feet of the couches, in the lower equipment bay, there is enough room to stand up.

When Apollo 11 lifted off, the spacecraft and launch vehicle combination stood 111 m (364 ft.) tall. Eight days later, when the flight ended, the only part recovered was the 3.3-m (11-ft.) tall *Columbia* command module.

Columbia was transferred to the National Air and Space Museum in 1970. It has been designated a Milestone of Flight by the museum and is prominently displayed near the 1903 Wright Flyer.

IN ORDER TO survive on the surface of the Moon, astronauts had to wear a suit that was self-contained, flexible, sturdy, and allowed them to perform activities requiring manual dexterity.

Among the many spacesuits in the collection of the National Air and Space Museum are the garments worn by the first two men to walk on the Moon, Neil Armstrong and Edwin "Buzz" Aldrin, on July 20, 1969.

These spacesuits were designed to provide a life-sustaining environment for the astronaut during periods of extravehicular activity in space or on the Moon or during unpressurized spacecraft operation. They could be worn with relative comfort for up to 115 hours in conjunction with the liquid cooling garment and be worn for fourteen days in an unpressurized mode.

Allowing for maximum mobility, each suit was custom fitted. The astronaut entered from the rear, through the pressure sealing slide fastener opening. Convoluted joint sections

DIMENSIONS: ══════

WEIGHT 81 KG (180 LB.)
ON EARTH;
13.6 KG (30 LB.)
ON THE MOON

1. The A-7L pressure suits worn by Neil Armstrong and Buzz Aldrin during the Apollo 11 lunar mission on display in the museum's Apollo to the Moon gallery. **2.** Astronaut Edwin E. "Buzz" Aldrin Jr., lunar module pilot, walks on the surface of the Moon near the leg of the lunar module *Eagle* on July 20, 1969. **3.** The crew of *Apollo 11,* Neil A. Armstrong, commander; Michael Collins, command module pilot; Edwin E. "Buzz" Aldrin Jr., lunar module pilot, pose in their spacesuits. **4.** An artist's cutaway Apollo A-7L spacesuit revealing the various layers and components of the suit.

of rubber were located in the shoulders, elbows, knees, hips, and ankles, enabling relatively easy movement while in the suit.

From the inside out, the suit was constructed of a nylon comfort layer, a neoprene-coated nylon pressure bladder, and nylon restraint layer. The outer layers consisted of Nomex and two layers of Teflon-coated Beta cloth, followed by layers of neoprene-coated nylon, layers of Beta/Kapton spacer laminate, and an outer layer of Teflon-coated Beta cloth. During the extravehicular activities (EVA), a liquid-cooling garment was worn closest to the skin, replacing the constant-wear garment.

The suits protected astronauts against micrometeoroids and temperatures ranging from −150°C to +120°C (−250°F to +230°F). The suits were made by the International Latex Corporation with the designation A-7L, and when combined with the portable life-support system and other components making up the extravehicular mobility unit, weighed approximately 82 kg (180 lb.) on Earth.

The pressure helmet was a transparent bubble designed to attach to the spacesuit neck ring. It was constructed of a transparent polycarbonate shell with a red anodized aluminum neck ring, a feed port, a vent pad at the rear, and a valsalva device attached to the inner ring.

The lunar extravehicular visor assembly (LEVA) consisted of a polycarbonate shell onto which the cover, visors, hinges, eyeshades, and latch were attached. There were two visors, one covered with a thermal control ultraviolet coating and the other with a gold optical coating. The two eyeshades could be raised or lowered to reduce low-angle solar glare. The LEVA was worn over the pressure helmet during EVA periods, and provided impact, micrometeoroid, thermal, and ultraviolet- and infrared-light protection.

The extravehicular gloves were built of an outer shell of Chromel-R fabric and thermal insulation to provide protection when handling extremely hot and cold objects. The blue fingertips were made of silicone rubber to provide more sensitivity.

The intravehicular gloves were made with a bladder dip molded from a cast of the individual's hand. The interior had an inner restraint core of nylon tricot that had been dipped in a neoprene compound. A convoluted section was incorporated into the wrist, with anodized aluminum connectors for attachment to the spacesuit. A fingerless glove restraint was attached to the bladder at the wrist and enclosed the entire hand excluding the fingers and thumb. The suits were transferred to the museum from NASA in 1971.

IN JULY 1975 two manned spacecraft—one launched from Kazakstan and the other from Florida—rendezvoused in orbit, fulfilling an agreement between the Soviet Union and the United States. This docking of an Apollo command and service module (CSM) and a Soyuz spacecraft was called the Apollo-Soyuz Test Project (ASTP).

Astronauts Thomas Stafford, Vance Brand, and Deke Slayton flew the mission, which launched on July 15, seven hours after the *Soyuz 19* was launched in the USSR with cosmonauts Alexei Leonov and Valeriy Kubasov aboard. Once docked, the Apollo and Soyuz crews visited each other's spacecraft, shared meals, and worked on various tasks and experiments over a two-day period. After separation, Apollo remained in space an additional six days and Stafford, Brand, and Slayton performed scientific experiments. Soyuz returned to Earth approximately forty-three hours after separation. The Apollo capsule splashed down on July 24, marking the end of all missions with Apollo spacecraft.

The workhorse of the Soviet and Russian space programs, the Soyuz ("union") has been in use longer than any other manned spacecraft. Designed in the 1960s, it first carried a cosmonaut into space in April 1967. Since then the Soyuz and its subsequent generations—the Soyuz T and Soyuz TM—have flown scores of manned missions in Earth orbit.

The Soyuz spacecraft has three main components: the large spherical section at the front is the orbital module; the landing module is the bell-shaped section in the middle; and the cylindrical section at the rear is the instrument module.

The rocket that launches the Soyuz-manned spacecraft is also called Soyuz. Used since 1963, the three-stage Soyuz launch vehicle has been modified several times. The Soyuz launch vehicle can loft a payload weighing up to 7,500 kg (16,500 lb.) into low Earth orbit. It is also used to launch scientific and military satellites.

1. A view of the Apollo-Soyuz Test Project (ASTP) display in the "Space Race" exhibit. 2. An official portrait of Apollo-Soyuz Test Project prime crew of the July 1975 Earth orbit mission. From left to right: astronauts Donald K. "Deke" Slayton, Thomas P. Stafford, Vance D. Brand, and cosmonauts Alexei A. Leonov and Valeriy N. Kubasov. 3. An historic handshake. Apollo commander, astronaut Thomas P. Stafford and Soyuz commander, cosmonaut Alexei A. Leonov make their historic handshake in the Universal Docking Adapter (UDA).

The Apollo-Soyuz Test Project marked a brief thaw in the Cold War and the first time that the two rival nations cooperated in a manned space mission. Engineering teams from both sides collaborated in the development of a common docking system that linked the two spacecraft. An airlock was needed to transition from the American cabin pressure system of 5 pounds per square inch of pure oxygen to the Soviet mixed oxygen/nitrogen system at normal atmospheric pressure (about 14.7 psi).

NASA contracted with North American Rockwell, the CSM contractor, to build the docking module (DM) quickly. On the front was mounted the three-leaf androgynous docking system. It could be used in either a passive (retracted) or active (extended) docking configuration.

The module in the National Museum of Air and Space is the backup DM, which was discarded a day before the Apollo reentered the atmosphere. The Block II Apollo command and service modules (CSM 105) on display in the Space Hall were originally used for vibration and acoustic tests. In 1973 they were refurbished for the Paris air show where they were docked to the Soviet Soyuz spacecraft. This exhibit was installed in NASM in 1976.

DIMENSIONS:

APOLLO COMMAND MODULE

LENGTH	3.7 M (12 FT.)
BASE DIAMETER	3.9 M (12 FT., 10 IN.)
WEIGHT	5,900 KG (13,000 LB.)

APOLLO SERVICE MODULE

LENGTH	6.7 M (22 FT.)
DIAMETER	3.9 M (12 FT., 10 IN.)
WEIGHT AT LAUNCH	6,800 KG (15,000 LB.)

APOLLO DOCKING MODULE

LENGTH	3 M (10 FT.)
DIAMETER	1.5 M (5 FT.)
WEIGHT	2,000 KG (4,400 LB.)

SOYUZ ORBITAL MODULE

LENGTH	2.5 M (8 FT., 2 IN.)
DIAMETER	2.2 M (7 FT., 2 IN.)
WEIGHT	1,200 KG (2,700 LB.)

SOYUZ LANDING MODEL

LENGTH	2.2 M (7 FT., 2 IN.)
DIAMETER	2.2. M (7 FT., 2 IN.)
WEIGHT	2,800 KG (6,200 LB.)

SOYUZ INSTRUMENT MODULE

LENGTH	2.3 M (7 FT., 6 IN.)
DIAMETER	2.2 M (7 FT. 2 IN.)
WEIGHT	2,650 KG (5,850 LB.)

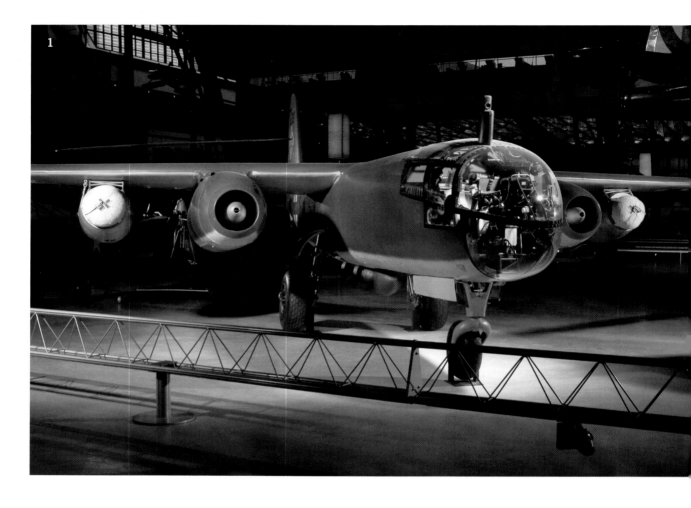

THE ARADO AR 234 B-2 Blitz ("lightning") was the world's first operational jet bomber and reconnaissance aircraft.

Development of the Ar 234 began in 1940 with the intention to build a reconnaissance aircraft propelled by the turbojet engines then under development by BMW and Junkers. Lead designer Rüdiger Kosin created a high-wing monoplane with two turbojet engines mounted in nacelles under the wings and two reconnaissance cameras in the rear fuselage.

The jet was redesigned for bombing capability. The fuselage was enlarged slightly to accommodate a conventional tricycle landing gear, and a semirecessed bomb bay was added under the fuselage to allow the pilot to act as a bombardier. There was a Lotfe 7K bombsight in the fuselage floor ahead of the control column, which the pilot swung out of his way to use as the sight.

The first prototype of the revised design, designated Ar 234 V9, flew on March 12, 1944. The bomber version, designated Ar 234 B-0, became the first subtype built in quantity. The German Air Ministry ordered 200 Ar 234 Bs. The initial order had called for two versions: the B-1 reconnaissance aircraft and the B-2 bomber.

There were plans for more advanced versions of the Arado jet, including the Ar 234 C powered by four BMW 003 A-1 engines and fitted with a pressurized cockpit. This was the fastest jet aircraft of World War II. However, only fourteen Ar 234 Cs left the Arado factory before Soviet forces overran the area.

Erich Sommer piloted the first Ar 234 combat mission on August 6, 1944, a reconnaissance sortie over the Allied beachhead in Normandy. He encountered no

DIMENSIONS:

WINGSPAN	14.44 M (47 FT., 3.5 IN.)
LENGTH	12.64 M (41 FT., 5.5 IN.)
HEIGHT	4.29 M (14 FT., 1.5 IN.)
WEIGHT	5,200 KG (11,464 LB.)

1. The museum's Arado 234, as displayed at the Udvar-Hazy Center.
2. A view from above pilot's seat of the cockpit of the National Air and Space Museum's Arado Ar 234. 3. An Arado Ar 234 B-2 (9V+CH) captured by the Royal Air Force at Sola airfield near Stavenger, Norway. 4. The museum's Arado Ar 234 following its capture by the Allied forces.

opposition and gathered more useful intelligence in two hours than the Luftwaffe obtained during the previous two months. Virtually immune to interception, the Ar 234 continued to provide the Germans with valuable reconnaissance until near the end of the war.

Only one Luftwaffe unit, KG 76 (Kampfgeschwader, or Bomber Wing, 76), was equipped with Ar 234 bombers. The unit flew its first missions during December 1944 in support of the Ardennes Offensive. Typical missions consisted of pinprick attacks conducted by fewer than twenty aircraft, each carrying a single 500-kg (1,100-lb.) bomb.

The unit participated in the desperate attacks against the Allied bridgehead over the Rhine at Remagen during mid-March 1945, but failed to drop the Ludendorff railway bridge and suffered a number of losses to antiaircraft fire. The deteriorating war situation, coupled with shortages of fuel and spare parts, prevented KG 76 from flying more than a handful of sorties. The unit conducted its last missions against Soviet forces encircling Berlin during the final days of April.

The unit's few surviving aircraft were either dispersed to airfields still in German hands or destroyed to prevent their capture. Acquired in the early 1950s, the National Air and Space Museum's Blitz, an Arado Ar 234 B-2 bomber, was one of nine Ar 234s surrendered to British forces at Sola airfield near Stavanger, Norway. It is the sole surviving example of an Ar 234.

LEONARD NIEMI'S SISU is the most successful American competition sailplane ever flown. John Ryan in 1962, Dean Svec in 1965, and A. J. Smith in 1967 all won the United States National Soaring Championships flying a Sisu (pronounced *see-soo*). In 1967, Bill Ivans set a national speed record flying a Sisu 1A at El Mirage, California, by skimming across the desert at 135 km/h (84 mph) over a 100-km (62-mi.) triangular course.

Alvin H. Parker flew from his hometown, Odessa, Texas, and set three world records: a declared-goal distance record flight of 784 km (487 mi.) in 1963, which Parker himself smashed in 1969 when he again reached his declared goal 931 km (578 mi.) from Odessa; and a free distance record of 1,042 km (647 mi.) in July 1964. This flight also shattered a symbolic and psychological 1,000-km barrier.

On August 3, 1967, Parker's fourteen-year-old son, Stephen, flew the Sisu 555 km (345 mi.) from Odessa to Farley, New Mexico, earning him the coveted Diamond C soaring badge for distance, altitude gain, and endurance flying. Stephen was the youngest American to earn the badge.

Aeronautical engineer Leonard A. Niemi started developing the Sisu 1 in 1952. Niemi

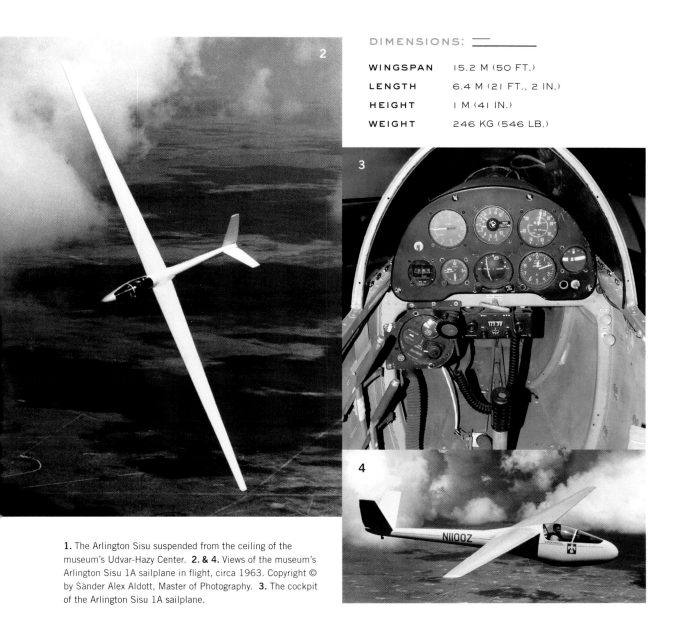

DIMENSIONS: ——

WINGSPAN	15.2 M (50 FT.)
LENGTH	6.4 M (21 FT., 2 IN.)
HEIGHT	1 M (41 IN.)
WEIGHT	246 KG (546 LB.)

1. The Arlington Sisu suspended from the ceiling of the museum's Udvar-Hazy Center. **2. & 4.** Views of the museum's Arlington Sisu 1A sailplane in flight, circa 1963. Copyright © by Sånder Alex Aldott, Master of Photography. **3.** The cockpit of the Arlington Sisu 1A sailplane.

chose to name his design the Sisu, a popular Finnish word with no precise English equivalent that describes a fundamental characteristic of native Finns referring to their strength, perseverance, and courage under adversity.

Niemi's Sisu flew on laminar-flow wings, and he designed it for private pilots to construct at home. The first flight in 1958 was so successful that Niemi decided not to sell plans or kits for the homebuilders but to modify the design for production as a finished, ready-to-fly sailplane. Some of his changes included lightening the wing structure, adding vents to the dive brakes, and introducing slotted flaps to expand the low-speed performance envelope even further. Niemi also increased slightly the area of the V-tail elevator/rudder to compensate for increased pitch force from the new slotted flaps, and increased the deflection range of the rudder too.

To manufacture the Sisu, Niemi set up the Arlington Aircraft Company about halfway between Dallas and Fort Worth, Texas. Construction began on the first four sailplanes in 1960. Pilots quickly snapped up these aircraft but production costs surpassed profits and Niemi had to sell the project. Six more Sisus were built, but profits never covered expenditures.

On July 12, 1967, at a ceremony atop Harris Hill close to Elmira, New York, the birthplace of American competitive soaring, Parker's record-breaking Sisu was formally presented to the Smithsonian.

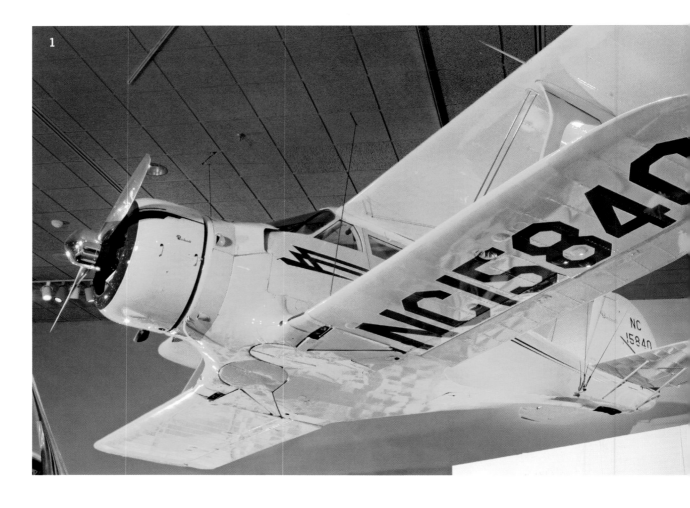

THE MODEL 17,

the first aircraft produced by the Beech Aircraft Company of Wichita, Kansas, was a gamble for founder and president Walter Beech, but it is considered a classic aircraft today.

Produced during the Depression, this expensive aircraft was designed as a high-speed, business airplane. The first series of Beech Staggerwings, of which only two were built, was known as the Model 17R. It had a fixed landing gear with wheel pants and was powered by a 420-hp Wright R-760 engine. It had a steel-tube fuselage and wing spar structure. The upper wing was inversely staggered behind the lower wing, and this design gave the Model 17 its unique shape and name.

The fuselage was faired with wood formers and stringers and was covered with fabric. Top speed of the aircraft was more than 322 km/h (200 mph) and yet the landing speed was only 97 to 105 km/h (60 to 65 mph). Such performance was remarkable for the time. Nevertheless, the aircraft was difficult to sell, largely because of its high cost.

In 1934, Beech introduced the Model B17L with retractable landing gear (which was not common at the time), wings of a different airfoil, the use of wood wing spars instead of steel ones, and a 225-hp Jacobs L-4 power plant. The improvements gave the aircraft a maximum speed of 282 km/h (175 mph) and a landing speed of 72 km/h (45 mph). The Beech Company advertised the performance and the airplane's gentle stall characteristics, both the result of the staggered-wing arrangement. The redesigned aircraft had hit the target, and sales improved significantly.

In all, 781 Beech 17s were produced in eight different series. The Staggerwing was not only a successful corporate aircraft, but a winner in racing circles, too. NC15835, a

DIMENSIONS:

WINGSPAN	9.75 M (32 FT.)
LENGTH	7.44 M (24 FT., 5 IN.)
HEIGHT	2.59 M (8 FT., 6 IN.)
WEIGHT	827.8 KG (1,825 LB.)

1. & 3. The museum's Beech C17L Staggerwing hanging on display in the Golden Age of Flight gallery. **2.** Beech D17S Staggerwing *Miss Streamline III* warming up its engine. **4.** A Beechcraft C17R with its engine running.

Model C17R, piloted by Louise Thaden and Blanche Noyes, won the 1936 Bendix Trophy Race, marking the first time that a woman had won that prestigious race. Other stock Staggerwings won two major air races in Miami in 1936. In 1937, Jacqueline Cochran set a 1,000-km speed record averaging more than 320 km/h (200 mph). Staggerwings also did well in the 1937 and 1938 Bendix Races.

The Beechcraft Staggerwing in the National Air and Space Museum was manufactured on July 3, 1936. The original owner of the aircraft was E. E. Aldrin, the father of astronaut Buzz Aldrin. It had nineteen different owners before it was donated to the museum in 1981 by Desert Air Parts of Tucson, Arizona.

The fact that this aircraft has survived so long is a testimony to its masterful design. Technologically advanced for its time, it will always remain a classic beauty.

WHAT THE JEEP was to American soldiers during World War II, so was the Huey helicopter to those who fought in Vietnam. All branches of the US military operated them in every corner of South Vietnam, Cambodia, and Laos.

"Huey" originated as a derivative of the original designation HU-1A (Helicopter, Utility, Model 1A). For a time, the Huey was one of the most recognizable aircraft in history due to the unmistakable *whop whop whop* of the main rotor blade.

The concept for this aircraft sprang from the Korean War, where the original Mobile Army Surgical Hospital (MASH) helicopter, the Bell 47 (H-13), recovered thousands of wounded soldiers and delivered them straight to critical care units. In 1954 the US Army launched a design competition for a new medical evacuation (medevac) helicopter.

The Lycoming Company was developing the XT-53 engine with army backing. There was no specific application for the engine, but Bell engineers saw great potential in this power plant and joined with Lycoming. Tests from the prototype helicopter they developed, the XH-40, performed so well that Bell earned a contract to produce 200 production medevac versions plus 100 outfitted as instrument trainers.

Designated HU-1A, it was officially christened the "Iroquois," in keeping with the army tradition of naming helicopters after American Indian tribes. In 1962 the army aircraft designation system changed, and the HU-1 became the UH-1, but the Huey nickname remained.

DIMENSIONS:

ROTOR	
DIAMETER	14.6 M (48 FT.)
LENGTH	12.8 M (42 FT.)
HEIGHT	4.4 M (14 FT., 5 IN.)
WEIGHT	2,293 KG (5,055 LB.) EMPTY

1. The museum's Bell UH-1H Iroquois (Huey) on display at the Udvar-Hazy Center. **2.** A view of the twin M-60 machine guns in the door of NASM's *Smokey III* in Vietnam, circa September 1967 through June 1968. **3.** A line of four US Army Bell UH-1H Hueys deploying troops along a dirt road somewhere in Vietnam. **4.** Four US Army soldiers seated in a Bell UH-1H Iroquois (Huey) helicopter on the ground in Vietnam.

The first Hueys to operate in Vietnam were medevac HU-1As that arrived in April 1962. These Hueys supported the South Vietnamese Army, but American crews flew them. In October the first armed Hueys equipped with 2.75-inch rockets and .30-caliber machine guns began flying in Vietnam. The main role of these Hueys was to escort army and marine transport helicopters. By the end of 1964, the army was flying more than 300 Hueys.

During the next decade, the Huey was upgraded and modified based on lessons learned in combat. In September 1967 the army accepted its first UH-1H Huey with a more powerful 1,400-hp Lycoming T53-L-13 engine, which had enough power to handle almost any mission in the harsh conditions of Vietnam. The Huey mission portfolio now covered troop transport, medevac, gunship, smoke ship, command and control, general service and support, and reconnaissance.

The Hueys profoundly affected the survival rate of battlefield casualties. Hueys airlifted 90 percent of these casualties directly to medical facilities, and the percentage of soldiers who died from wounds sustained in combat fell to 19 percent, about a 25 percent drop from the Korean War fatality rate.

The National Air and Space Museum acquired a UH-1H Huey in 1995. Before that, this aircraft did four tours in Vietnam and had more than 2,500 combat flying hours. It is now on public display at the Steven F. Udvar-Hazy Center.

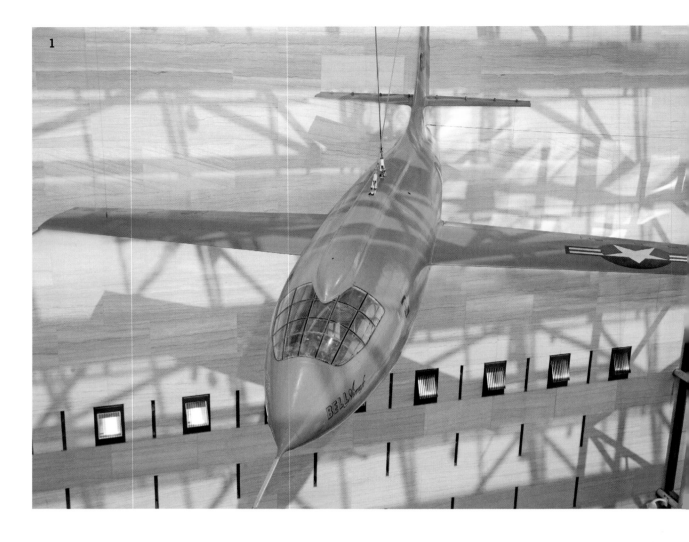

1

ON OCTOBER 14, 1947, flying the Bell X-1 #1, Capt. Charles E. "Chuck" Yeager, USAF, became the first pilot to break the sound barrier.

The X-1 was developed as part of a cooperative program initiated in 1944 by the National Advisory Committee for Aeronautics (NACA) and the US Army Air Forces (later the US Air Force) to develop special manned transonic and supersonic research aircraft. This was necessary because, by the end of World War II, many aircraft were encountering extreme buffeting at high speeds that often caused a loss of control or even the destruction of the airframe. The buffeting was caused by shock waves in front of the aircraft as it approached the speed of sound.

The Army originally assigned the designation XS-1 for Experimental Sonic-1, and three aircraft were built. The National Air and Space Museum owns the X-1 #1, named *Glamorous Glennis* by Captain Yeager in honor of his wife. The X-1 #2 was flight-tested by NACA. The X-1 #3 had a turbopump-driven, low-pressure fuel feed system. This aircraft was lost in a 1951 explosion on the ground. Three additional X-1 aircraft, the X-1A, X-1B, and X-1D, were constructed and test-flown.

The two X-1 aircraft were constructed from high-strength aluminum, with propellant tanks fabricated from steel. The first two X-1 aircraft did not utilize turbopumps for fuel feed to the rocket engine, relying instead on direct nitrogen pressurization of the fuel-feed system. The smooth contours of the X-1, patterned on the lines of a .50-caliber machine-

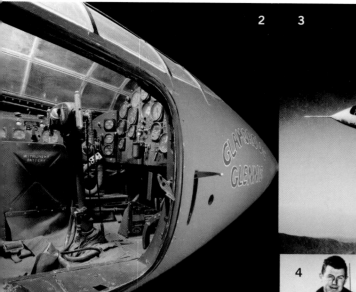

1. The Bell X-1 *Glamorous Glennis* on display in the National Air and Space Museum Milestones of Flight gallery. **2.** The cockpit of Bell X-1 *Glamorous Glennis* while on display at the National Air and Space Museum. **3.** The Bell X-1 *Glamorous Glennis* in flight over Muroc Dry Lake in California, flown by Chuck Yeager. Note the shock diamond pattern in the exhaust plume. **4.** Charles "Chuck" Yeager poses in his leather jacket beside of nose of *Glamorous Glennis*.

DIMENSIONS:	
WINGSPAN	8.53 M (28 FT.)
LENGTH	9.41 M (30.9 FT.)
HEIGHT	3.31 M (10.85 FT.)
WEIGHT	3,175 KG (7,000 LB.)

gun bullet, masked an extremely crowded fuselage containing two propellant tanks, twelve nitrogen spheres for fuel and cabin pressurization, the pilot's pressurized cockpit, three pressure regulators, retractable landing gear, the wing carry-through structure, a Reaction Motors, Inc., 6,000-pound-thrust rocket engine, and more than 227 kg (500 lb.) of special flight-test instrumentation.

Though originally designed for conventional ground takeoffs, all X-1 aircraft were air-launched from Boeing B-29 (see page 28) or B-50 Superfortress aircraft. Performance and safety hazards associated with operating rocket-propelled aircraft from the ground caused mission planners to resort to air-launching. Nevertheless, on January 5, 1949, *Glamorous Glennis,* piloted by Chuck Yeager, successfully completed a ground takeoff from Muroc Dry Lake in California. The maximum speed attained by the X-1 #1 was Mach 1.45 at 12,232 m (40,130 ft.), approximately 1,540 km/h (957 mph), during

a flight by Yeager on March 26, 1948. On August 8, 1949, Maj. Frank K. Everest Jr., USAF, reached an altitude of 21,916 m (71,902 ft.), the highest flight made by the rocket airplane. It continued flight-test operations until mid-1950, completing a total of nineteen contractor demonstration flights and fifty-nine air force test flights.

On August 26, 1950, Air Force Chief of Staff Gen. Hoyt Vandenberg presented the X-1 #1 to Alexander Wetmore, then secretary of the Smithsonian. The X-1, General Vandenberg stated, "marked the end of the first great period of the air age, and the beginning of the second. In a few moments the subsonic period became history and the supersonic period was born." Earlier, Bell Aircraft president Lawrence D. Bell, NACA scientist John Stack, and US Air Force test pilot Chuck Yeager had received the 1947 Robert J. Collier Trophy for their roles in first exceeding the speed of sound and opening the pathway to practical supersonic flight.

1

THOUGH THE FIRST American jet aircraft did not see combat, the Bell XP-59A Airacomet provided valuable technology that paved the way to more advanced designs.

In February 1941, Gen. Henry H. "Hap" Arnold, deputy chief of staff for air, saw the British Gloster E.28/39 jet-propelled test airplane fly in England. A W.1X turbojet engine designed by Frank Whittle powered the E.28/39. After exchanging information with the British, and in addition to turbine engine research already under way in the United States, the Army Air Forces and General Electric met and decided that the United States must begin at once to construct a jet engine based on the new Whittle W.2B engine. General Electric would build the engines, and Bell Aircraft Corporation would build the airframe.

In September 1942, Bell shipped the first XP-59A to a remote base in California for the initial flight trials. To maintain secrecy, Bell mounted a dummy propeller on the nose and threw a tarpaulin over the fuselage to disguise the Airacomet as just another new piston engine aircraft. Mechanics removed the propeller before flight and reinstalled it after the airplane landed. On October 1, Bell test pilot Robert M. Stanley took the XP-59A into the air for the first time. During this initial flight, Stanley kept the landing gear fully extended and flew no higher than 7.6 m (25 ft.). Later that day, he made three more flights and reached heights of 30 m (100 ft.). By the next day the plane was flying as high as 3,048 m (10,000 ft.).

DIMENSIONS: ⎯⎯

WINGSPAN	14.93 M (49 FT.)
LENGTH	11.83 M (38 FT., 10 IN.)
HEIGHT	3.76 M (12 FT., 3.75 IN.)
WEIGHT	3,320 KG (7,320 LB.)

1. NASM's Bell XP-59A Airacomet hanging on display in the Milestones of Flight gallery. It is the very first US jet ever flown. **2.** A Bell XP-59A Airacomet in flight in 1943. **3.** A Bell XP-59A Airacomet on the ground, with technicians underneath working on the aircraft.

Two General Electric Type I-A centrifugal-flow jet engines drove the XP-59A airframe to a maximum speed of only 628 km/h (390 mph). A number of enemy and Allied piston-engine fighters exceeded this velocity, so in March 1942 Bell received a contract for thirteen YP-59A test and evaluation aircraft with more powerful General Electric I-16 (J31) turbojet engines, which powered these and all subsequent production Airacomets.

The first of the YP-59As were flight-tested in June 1943. One of these aircraft set an unofficial altitude record of 14,512 m (47,600 ft.). Eventually, Bell completed only fifty production Airacomets—twenty P-59As and thirty P-59Bs. Each was armed with one 37-mm M-4 cannon and three .50-caliber machine guns. The P-59Bs were assigned to the 412th Fighter Group. The P-59 aircraft could fly at a maximum speed of 658 km/h (409 mph) at 10,640 m (35,000 ft.).

America's first XP-59A had amassed only 59 hours and 55 minutes of flying time before it came to the Smithsonian in 1945. The staff restored the plane to its original configuration for the opening of the new National Air and Space Museum in 1976. Today, the Airacomet hangs in the Milestones of Flight gallery.

1

LOUIS BLÉRIOT, a French engineer and manufacturer of automobile headlamps and other accessories, first became interested in aeronautics in 1901 when he constructed an experimental ornithopter. During the next eight years, he moved through a series of ten distinct aircraft designs, only one of which was capable of making a flight of more than ten minutes.

The next effort, the Type XI, was designed primarily by engineer Raymond Saulnier, but it was a natural evolution from his earlier aircraft and one to which Blériot himself made substantial contributions. It was first flown at Issy-les-Moulineaux, a military parade ground converted into a flying field, on January 23, 1909. In May the Type XI was fitted with a crude but reliable 25-hp, three-cylinder Anzani engine, and flown with regular success.

Blériot achieved immortality in the Type XI on July 25, 1909, when he made the first airplane crossing of the English Channel, covering the 40 km (25 mi.) between Calais and Dover in 36 minutes, 30 seconds. It was not the longest flight to date, either in duration or distance, but the symbolic impact of conquering the Channel by airplane made it the most widely acclaimed flight before that of Lindbergh. For his effort, Blériot captured the *London Daily Mail* prize of $2,500 that had been put up by the newspaper the year before for any successful cross-Channel airplane flight.

After the Channel flight, Blériot received many orders for his Type XI. Variants of the original 1909 machine were produced by the Blériot firm, foreign licensees, and enthusiastic amateur builders in Europe and America into World War I. Hundreds were built, many for

DIMENSIONS: ⎯⎯⎯

WINGSPAN	8.52 M (28 FT., 6 IN.)
LENGTH	7.63 M (25 FT., 6 IN.)
HEIGHT	2.7 M (8 FT., 10 IN.)
WEIGHT	326 KG (720 LB.)

1. The museum's restored Blériot XI in front of the restoration building at the Paul E. Garber Facility. **2.** A ground crew holds the tail of a Blériot XI at an early air meet. **3.** *Blériot Crosses the Channel;* an artist's rendition of Louis Blériot's first successful crossing of the English Channel by air, July 25, 1909. **4.** Louis Blériot at the controls of his Blériot XI in which he crossed the English Channel.

leading aviators of the day. Blériot's original Type XI today is in the possession of the Conservatoire des Arts et Métiers in Paris.

The Blériot XI in the National Air and Space Museum was manufactured by Blériot Aéronautique at Levallois-Perret, France, in 1914. It is a standard Type XI of the immediate prewar period, powered by a 50-hp Gnôme rotary engine with wing-warping for lateral control and a castering undercarriage that eased the problem of crosswind landings. The airplane was purchased by the Swiss aviator John Domenjoz, a Blériot company flight instructor. Domenjoz had recently earned a reputation as one of Europe's most celebrated stunt pilots while performing in major European cities during the early summer of 1914. Planning to continue his European exhibition tour, Domenjoz ordered this airplane to be specially strengthened and added a heavy harness to support the pilot during inverted flight.

With the outbreak of war in Europe, Domenjoz took his new Blériot to South America, where he continued to thrill crowds with daring aerobatics. He often flew inverted for extended periods of time, and flying at Buenos Aires in April 1915, he performed forty consecutive loops in 28 minutes. Such feats earned him the nickname "Upside-down Domenjoz." He returned to the United States in the fall of 1915 and made headlines flying in and around New York City. Domenjoz and his Blériot continued the tour through the southern and midwestern states in late 1915 and early 1916. Then he took his show to Havana, Cuba. Domenjoz made one final barnstorming tour in the United States in 1919 before placing the Blériot in storage on a Long Island farm and returned to France, where he remained for seventeen years.

The Blériot was sold to a museum on Long Island to cover unpaid storage costs. In 1950 the Smithsonian purchased it and then fully restored it by 1979. The Blériot Type XI is now part of the National Air and Space Museum's Early Flight gallery.

1

THE FIRST BOEING 247 made its initial flight on February 8, 1933, and the plane's performance confirmed the wisdom of Boeing management. Three key men—president Phillip G. Johnson, vice president Claire Egtvedt, and chief engineer C. N. Monteith—chose to develop the transport potential of their Boeing B-9 twin-engine bomber rather than stick to the trimotor and biplane design of the day.

The group of United predecessors (Boeing Air Transport, Pacific Air Transport, National Air Transport, and Varney Air Lines) replaced its entire fleet by ordering sixty 247s, making all other transports obsolete overnight and gaining advantage over the competitors.

The all-metal, low-wing 247 combined a retractable landing gear, two supercharged air-cooled engines, and, in later models, controllable-pitch propellers, with totally new standards in passenger comfort. The ten passengers and three crew members enjoyed excellent soundproofing, a low vibration level, and plush seats.

On May 22, 1933, the new 247 flew from San Francisco to New York in 19½ hours, compared with the previous time of 27 hours. Curiously, the inability of other airlines to obtain the 247 worked to Boeing's and United's net disadvantage. Transcontinental and Western Air (TWA) went to Douglas for a competitive aircraft, and the result was the famous DC series, which made the 247, in turn, obsolete.

The original 247 had a top speed of 293 km/h (182 mph) and cruised at 274 km/h (170 mph) compared with the 185 km/h (115 mph) of the Ford Tri-Motor (see page 74)

DIMENSIONS: ════

WINGSPAN	22.55 M (74 FT.)
LENGTH	15.72 M (51 FT., 7 IN.)
HEIGHT	3.7 M (12 FT., 1.75 IN.)
WEIGHT	4,148 KG (9,144 LB.)

1. The museum's Boeing 247-D hanging on public display in the Hall of Air Transportation. **2.** Pilots Roscoe Turner (left) and Clyde Pangborn (right) pose standing beside the museum's Boeing 247-D they flew as competitors in the MacRobertson Race from Mildenhall, England, to Melbourne, Australia, in October 1934. **3.** Passengers waiting to board the museum's 247-D in the mid-1930s. Note artwork aft of cabin door depicting route this aircraft took when it placed third in the 1934 MacRobertson Race.

then in general use. Boeing attempted to match the Douglas aircraft by creating the 247-D, with a 322 km/h (200 mph) top speed and 304 km/h (189 mph) cruise speed. Earlier 247s were modified to 247-D standards, but the airplane did not have the necessary growth potential to compete and was soon relegated to shorter route segments and smaller airlines.

The 247-D that is in the National Museum of Air and Space was leased from United by famed air racer Roscoe Turner and modified with extra fuel tanks to provide a range of more than 4,023 km (2,500 mi.) for the 1934 MacRobertson Race. Turner, Clyde Pangborn, and Reeder Nichols took off from Mildenhall, England, on October 20, 1934, and landed 92 hours, 55 minutes, and 30 seconds later in Melbourne, Australia, finishing in third place. The race was won by an English de Havilland DH 88 Comet, and second place went to a KLM-operated Douglas DC-2. Turner had an actual flying time of a little over 85 hours for the 18,186-km (11,300-mi.) distance and

might have finished second were it not for some engine problems and a navigational error that cost three hours' flying time.

The airplane was returned to United and served in regular airline service until 1937, when it was sold to the Union Electric Company of St. Louis for use as an executive transport. In 1939 it was purchased by the Department of Commerce Air Safety Board, which used it for fourteen years before presenting it to the museum in 1953. The aircraft served so well in so many experiments that it received the affectionate nickname "Adaptable Annie."

In 1974, United made a grant to have the 247-D restored for the museum. To highlight the most interesting aspects of the 247-D's career, the airplane is displayed with two sets of markings. The left side is marked as it was when flown by Turner with the NR-257Y registration; the right side is marked as it was when flown by United with the NC 13369 registration.

IN THE COMMERCIAL field, Boeing had led the world with their Model 247, which dominated aviation in 1933 but was soon supplanted by better, more efficient DC-2 and DC-3 designs from Douglas. In response, Boeing hoped to regain the lead through technology. And lessons from the Army Air Corps' Lockheed XC-35—the first aircraft with a pressurized fuselage—clearly showed the way.

Boeing designed a large, pressurized four-engine transport based on the airframe of a B-17C. Capable of cruising above 5,182 m (17,000 ft.), the new Boeing 307 Stratoliner was faster and far more comfortable than conventional aircraft, because it flew above most adverse weather. Construction began in 1937, after orders were received for ten aircraft from Pan American Airways and Transcontinental and Western Air (TWA).

When the 307 was completed late in 1938, Boeing had produced a stunning design. The 307 was equipped with a wide, comfortable fuselage that accommodated thirty-three passengers in unprecedented comfort, including collapsible beds, with a maximum range of 3,863 km (2,400 mi.) and a top speed of 396 km/h (246 mph). After the loss of the prototype during testing, Boeing redesigned the tail fin and rudder and installed it on all the subsequent 307s. The Stratoliner entered service with TWA on July 8, 1940.

Flying between New York and Los Angeles, TWA's five Boeing 307s crossed the country in less than 14 hours—two to four hours faster than the DC-3—stopping twice along the way to refuel. Pan American operated its three Stratoliners in 1940 from Miami to destinations throughout South America, replacing Pan Am's venerable flying boats on most of these routes. Pan Am also operated Stratoliners from Los Angeles and Brownsville, Texas, to points in Mexico and Central America.

1. The museum's Pan American Airways Boeing S-307 (PAA-307) Stratoliner *Clipper Flying Cloud* on display at the Udvar-Hazy Center. **2.** A Pan American Airways copilot speaks with the flight engineer seated behind him in the cockpit of a Boeing 307 Stratoliner. **3.** The museum's Stratoliner in flight. **4.** A Trans World Airlines (TWA) brochure emphasizes the high level of service on the 307.

DIMENSIONS:

WINGSPAN	32.7 M (107 FT., 3 IN.)
LENGTH	22.7 M (74 FT., 4 IN.)
HEIGHT	6.3 M (20 FT., 9 IN.)
WEIGHT	13,748.6 KG (30,310 LB.)

The Stratoliner would have had a long and successful commercial career had the Second World War not intervened. In December 1941 most of the nation's airliners were transferred to the armed forces. Designated C-75s, TWA's 307s were stripped of their pressurization systems to save weight. They were then contracted to the US Army Air Forces and flown by TWA crews on military transport missions throughout the war. The unpressurized 307s were returned to TWA and refurbished for commercial passenger use by spring 1945.

Pan American flew its 307s unmodified throughout the war on many military missions but with Pan Am flight crews. After the war these Stratoliners flew between New York and Bermuda until they were sold in 1946.

TWA retired its 307s in 1950 and sold them to Aigle Azur, a French airline flying in Europe and in Southeast Asia. Operating during the Vietnam War for Compagnie Internationale de Transports Civils Aérien (CIC), two TWA and one Pan Am Stratoliner flew regularly scheduled service for the United Nations between Hanoi and Saigon during hostilities.

One Stratoliner was purchased in 1939 by Howard Hughes for his personal use and for an around-the-world flight. When World War II prevented Hughes from attempting his flight, he converted the aircraft into a "Flying Penthouse" equipped with a luxurious interior in which he entertained numerous Hollywood friends and associates. It was scrapped and its fuselage converted into a posh houseboat known as "the Cosmic Muffin."

The National Air and Space Museum's Stratoliner started life as Pan American's *Clipper Flying Cloud*. After the war it was sold to the Airline Training Company and resold in 1954 to the Army Air Corps of Haiti. During this time it flew for the Compagnie Haitienne de Transports Aériens and often served as the personal aircraft of Haitian president and dictator François "Papa Doc" Duvalier. After 1957 it was sold to a succession of private owners until it was purchased in 1969 by Aviation Specialties of Mesa, Arizona, for conversion into a firebomber.

The museum acquired the 307 in 1972. In 1994 the Boeing Company generously offered to restore the aircraft, and after nine years of dedicated work the Boeing 307 was flown to its new home in the Steven F. Udvar-Hazy Center in August 2003.

THE BOEING 367-80, also known as the Dash 80, revolutionized commercial air transportation as America's first jet airliner.

The 367-80 was based on the C-97, a piston-engine transport and aerial tanker intended for use by the US Air Force. By the time Boeing progressed to the eightieth iteration, the design bore no resemblance to the C-97 and became known as the 367-80.

The 367-80 mated the large cabin from the C-97 with the 35-degree swept-wing design of the B-47 and B-52 but was considerably stiffer and had pronounced dihedral. The wings were mounted low on the fuselage and incorporated high-speed and low-speed ailerons as well as a sophisticated flap and spoiler system. Four Pratt & Whitney JT3 turbojet engines, each capable of producing 4,536 kg (10,000 lb.) of thrust, were mounted on struts beneath the wings.

On its first flight on July 15, 1954, the 367-80 flew 161 km/h (100 mph) faster than the de Havilland Comet. It also had a maximum range of more than 5,633 km (3,500 mi.). The air force convinced Boeing to widen the design by .3 m (12 in.) and eventually bought over seven hundred. They designated theirs the KC-135A.

2

DIMENSIONS:

WINGSPAN	39.5 M (129 FT., 8 IN.)
LENGTH	39 M (127 FT., 10 IN.)
HEIGHT	11.6 M (38 FT.)
WEIGHT	41,785 KG (92,120 LB. EMPTY)

3

4

1. The Boeing 367-80 on display at the Udvar-Hazy Center. **2.** The Boeing 367-80 in flight over broken cloud layer shortly after its first flight in July 1954. **3.** A. M. "Tex" Johnson (left), Boeing's chief of flight test, and Richards L. "Dix" Loesch examine the flight engineer's instrument panel of the Boeing Model 367-80. **4.** A Pan American Boeing 707-321C in flight.

Boeing turned their attention to selling 367-80s to the airline industry. At the Gold Cup hydroplane races held on Lake Washington in Seattle in August 1955, test pilot Alvin "Tex" Johnston barrel-rolled the 367-80 over the lake in full view of thousands of astonished spectators and airline representatives.

Pan American Airways president Juan Trippe was looking for a new jet airliner to enable his pioneering company to maintain its leadership in air travel. The 367-80, now known as the 707, was widened to seat six passengers per row rather than five. Pan Am ordered twenty-five. Now able to carry 160 passengers, the 707 fuselage became the standard design for all of Boeing's subsequent narrow-body airliners.

In October 1958, Pan Am opened international service with the Boeing 707. National Airlines inaugurated domestic jet service two months later using a 707-120 borrowed from Pan Am. American Airlines flew the first domestic 707 jet service with its own aircraft in January 1959. Nonstop flights between New York and San Francisco took only five hours—three hours less than by piston-engine DC-7. The flight was almost 40 percent faster and almost 25 percent cheaper.

Boeing built 855 707s. Until its retirement in 1972, the 367-80 tested numerous advanced systems, including a fifth engine mounted on the rear fuselage.

The Boeing 367-80 was donated to the Smithsonian in 1972. It was restored by Boeing in the 1990s and flown to the Steven F. Udvar-Hazy Center in August 2003.

BOEING'S B-29 SUPERFORTRESS was the most sophisticated propeller-driven bomber to fly during World War II, and the first bomber to house its crew in pressurized compartments.

During the war in the Pacific Theater, B-29s destroyed most of Japan's cities with conventional bombs and delivered the first nuclear weapons used in combat. On August 6, 1945, Col. Paul W. Tibbets Jr., in command of the Superfortress *Enola Gay,* dropped an atomic bomb on Hiroshima, Japan. Three days later, another B-29, named *Bockscar,* dropped a second atomic bomb on Nagasaki, Japan. On August 15, 1945, the Japanese accepted Allied terms for unconditional surrender.

As war clouds darkened Europe in the late 1930s, the US Army Air Corps realized that their B-17 Flying Fortress and B-24 Liberator would not be enough. Boeing responded with the B-29 bomber, which could carry a maximum bomb load of 909 kg (2,000 lb.) at a speed of 644 km/h (365 mph) a distance of 8,050 km (5,000 mi.).

The plane had a long, narrow, high-aspect ratio wing equipped with large Fowler-type flaps. This wing design allowed the B-29 to fly fast at high altitudes without also having problems during the slower speeds required for landing and takeoff. The flight deck forward of the wing, the gunner's compartment aft of the wing, and the tail gunner's station were all pressurized. For the crew, flying at extreme altitudes was almost comfortable.

To protect the Superfortress, there were five unmanned gun turrets on the fuselage: one above and behind the cockpit that housed two .50-caliber machine guns, another aft

1. The nose and inboard engines of the Boeing B-29 Superfortress *Enola Gay* seen on the floor at the Udvar-Hazy Center. **2.** Unidentified men pose beneath the nose of the Boeing B-29 Superfortress *Enola Gay* on Tinian, Mariana Islands, 1945. **3.** The *Enola Gay* landing at Tinian, Marianas Islands, after completing mission to drop atomic bomb on Hiroshima, Japan, August 6, 1945. **4.** Lt. Col. John Porter with flight crew of the *Enola Gay:* Capt. Theodore J. "Dutch" Van Kirk, navigator; Major Thomas W. Ferebee, bombardier; Col. Paul W. Tibbets, pilot; Capt. Robert A. Lewis, copilot; and Lt. Jacob Beser, radar countermeasure officer. Sgt. Joseph S. Stiborik, radar operator; S/Sgt. George R. Caron, tail gunner; Pfc. Richard H. Nelson, radio operator; Sgt. Robert H. Shumard, assistant engineer; and S/Sgt. Wyatt E. Duzenbury, flight engineer.

near the vertical tail equipped with two machine guns, plus two more beneath the fuselage, each equipped with two .50-caliber guns. Gunners operated these turrets by remote control by aiming through computerized sights. Another two .50-caliber machine guns and a 20-mm cannon were fitted in the tail beneath the rudder and fired manually.

Boeing also equipped the B-29 with advanced radar equipment and avionics. Depending on the type of mission, a B-29 carried the AN/APQ-13 or AN/APQ-7 Eagle radar system to aid bombing and navigation. These systems were accurate enough to permit blind bombing through clouds that completely obscured the target. The B-29B was equipped with the AN/APG-15B airborne radar-gun sighting system mounted in the tail, ensuring accurate defense against enemy fighters attacking at night.

By May 1944 there were 130 B-29s in the 20th Air Force operational from airfields in India. On June 5, 1944, the B-29 flew its first combat mission against Japanese targets in Bangkok, Thailand.

With the fall of Saipan, Tinian, and Guam in the Mariana Islands chain in August, the US Army Air Forces (AAF) acquired air bases that lay several hundred miles closer to Japan. A strike with hundreds of B-29s was launched at night from low altitude, using incendiary bombs that devastated much of Japan.

Late in 1944 the AAF modified fifteen B-29s by deleting all gun turrets, except for the tail position, removing armor plate, installing Curtiss electric propellers, and configuring the bomb bay to accommodate atomic weapons. The *Enola Gay* was assigned to the 509th Composite Group commanded by Colonel Tibbets, who named B-29 #82 after his mother. The *Enola Gay* dropped the atomic bomb "Little Boy" over Hiroshima on August 6, 1945.

The *Enola Gay* was transferred to the Smithsonian on July 4, 1949. The plane's restoration, the largest such project ever undertaken at the National Air and Space Museum, took more than twenty years to complete.

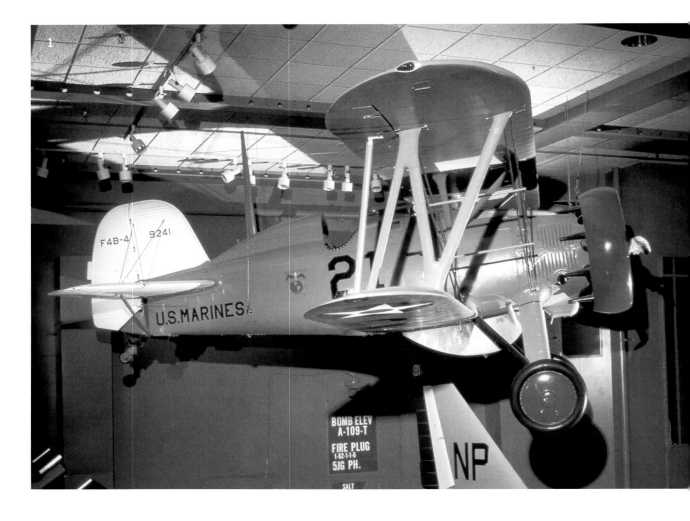

THE LAST WOODEN-WINGED biplane designed for the military, the Boeing F4B/P-12 series served as the primary fighter of the US Navy and Army Air Corps from the early 1930s through the early 1940s.

Two prototypes designated XF4B-1 were first delivered to the navy for evaluation in 1928. Convinced of the merits of the design after extensive trials, the navy purchased twenty-seven production aircraft. Sent to the USS *Lexington* in the summer of 1929, the new fighter was capable of reaching speeds of more than 282 km/h (175 mph) and could carry five 11-kg (24-lb.) bombs under each wing, with either one 225-kg (500-lb.) bomb or one 155-l (41-gal.) fuel tank beneath the fuselage. Armament on the F4B-1 consisted of two .30-caliber machine guns synchronized to fire through the propeller arc.

Following the success of the first model, the navy contracted for forty-six improved versions. The F4B-2 differed in having a redesigned ring cowling, improved split-axle landing gear, and Friese ailerons. Maximum speed was increased to 298 km/h (186 mph), and the airplane could carry four 56.2-kg (116-lb.) bombs.

Encouraged by the navy's results with the XF4B-1, the army placed an order for ten similar aircraft with the carrier hook deleted. The army version was designated the P-12. The next model, the P-12B, was upgraded with Friese ailerons and a shorter landing gear. Ninety were produced. In June 1930 the Army Air Corps contracted for 131 P-12C models, which incorporated the P-12B airframe with a ring cowl and a cross-axle landing gear.

While production of the F4B-2 was in progress, Boeing began development of a new version. Instead of the bolted, alloy-tube fuselage of the earlier design, the F4B-3 had an all-

DIMENSIONS:

WINGSPAN	9.2 M (30 FT., 2 IN.)
LENGTH	6.2 M (20 FT., 4 IN.)
HEIGHT	2.9 M (9 FT., 6 IN.)
WEIGHT	1,070 KG (2,354 LB.)

1. The museum's US Marine Corps Boeing F4B-4 hanging on display in the Sea-Air Operations gallery. **2.** A period shot of two unidentified Marine mechanics affixing practice bombs to the museum's Boeing F4B-4. **3.** A view of a formation of three US Navy Boeing F4B-4 (Model 235) aircraft in flight.

metal semimonocoque fuselage and a Pratt & Whitney R-1340-10, which was fitted with a drag ring. The navy contracted for seventy-five F4B-3s. The army ordered 135 P-12Es, and twenty-five were delivered as P-12Fs, which had Pratt & Whitney SR-1340G engines for increased high-altitude performance. The fourth and final version of the F4B series, which had a broader chord fin and a larger

headrest that housed an inflatable life raft, was the F4B-4. Twenty-one of these airplanes were assigned to the US Marine Corps.

The airplane in the National Air and Space Museum is one of the F4B-4s built for the marines. It was donated to the museum in 1959, and is now in its former colors as airplane number 21 of Marine Fighting Squadron VF-9M.

KNOWN AFFECTIONATELY AS the "Peashooter," the Boeing P-26 fighter represents a turning point in US military aircraft design; it introduced the concept of the high-performance, all-metal monoplane fighter. While it was a radical departure from the wood-and-fabric biplane, the P-26 retained features of its predecessors such an open cockpit, fixed landing gear, and an external wing bracing.

Design of the P-26 started in September 1931 as a joint Boeing and US Army project, incorporating features proposed by both parties. Boeing constructed the airframe and the army provided the engines, instruments, and other necessary equipment. Based on the success of the first prototype flown in March 1932, the army contracted for 111 improved production aircraft designated P-26A.

The structure of the P-26 was based to a great extent on experience gained in creating the Boeing Monomail and other Boeing all-metal designs. However, unlike the Monomail, the Peashooter did not have cantilevered wings or retractable landing gear, opting for the lighter structure that external bracing allowed. The fixed gear produced considerable drag, but it greatly reduced weight and structural complexity.

The fuselage was of semimonocoque construction with aluminum bulkheads, formers, longerons, and skin. The wings were built of duralumin with two main spars supporting the ribs and the skin, which were braced with external steel wires. The fully cantilevered tail surfaces were of single spar, metal skin construction.

2

DIMENSIONS:

WINGSPAN	8.5 M (27 FT., 11 IN.)
LENGTH	7.3 M (23 FT., 11 IN.)
HEIGHT	3.1 M (10 FT., 2 IN.)
WEIGHT	996 KG (2,196 LB.)

3

1. The museum's Boeing P-26A hanging on exhibit at the Udvar-Hazy Center. 2. The cockpit of the National Air and Space Museum's Boeing P-26A Peashooter. 3. A Boeing P-26 airplane in flight over Oahu, on March 6, 1939; it was part of the 19th Pursuit Squadron, 18th Pursuit Group.

Power was provided by a 600-hp, nine-cylinder air-cooled Pratt & Whitney R-1340-27 Wasp radial engine, enclosed in an NACA cowling ring. Maximum speed in level flight was 374 km/h (234 mph) at 2,300 m (7,545 ft.), with a service ceiling of 8,350 m (27,400 ft.). Armament consisted of either two .30-caliber machine guns or one .30-caliber and one .50-caliber gun, synchronized to fire through the propeller arc. Two 45-kg (100-lb.) or five 13.6-kg (30-lb.) bombs could also be carried.

Originally, P-26As were built with streamlined headrests. Following the death of an army pilot when his airplane overturned on landing, all P-26s were modified with a larger, stronger headrest.

After P-26A production had been completed, the army sought to reduce the inordinately high landing speed of the airplane by installing experimental flaps. The modification reduced the landing speed from 132 km/h (82.5 mph) to 117 km/h (73 mph). Boeing retrofitted the flaps to all A models and equipped the B and C models that were under construction. P-26Bs and Cs were identical to the original except for the fuel-injected R-1340-33 Wasp engine and modifications to the fuel system.

When the P-26 was removed from regular service, those aircraft stationed overseas were sold to the Philippines or were assigned to the Panama Canal Department Air Force (a branch of the US Army Air Corps). Eleven P-26As were sold to China and one to Spain. Those serving with China and the Philippines in the 1930s and 1940s fought gallantly against the invading Japanese, scoring numerous successes.

The P-26A in the National Air and Space Museum was first given to Guatemala in late 1942 and flew in their air force from 1943 to 1954. In 1957 the government of Guatemala donated it to the Smithsonian. The airplane was restored by the US Air Force for the Smithsonian and was displayed at the US Air Force Museum in Dayton, Ohio, until 1975, when it was brought to Washington, DC. It is painted in the colors of the 34th Attack Squadron that was stationed at March Field, California.

1

ON MARCH 21, 1999, Bertrand Piccard and Bryan Jones guided *Breitling Orbiter 3,* the first balloon to fly around the world nonstop, to a safe landing on a desolate stretch of desert in western Egypt.

Since 1980, eighteen teams had announced their intention to circumnavigate the globe in a balloon. Seven of those teams launched a total of sixteen balloons in unsuccessful attempts. In the fall of 1997 the Anheuser-Busch Company offered a trophy and a $1 million prize (half of which was to be donated to charity) to the first balloonists to achieve what was widely recognized as "the last great aviation challenge of the century." While none of the teams had achieved an around-the-world flight by the end of 1998, they had captured the public imagination with a series of record-breaking long-distance flights and hair breadth escapes from danger.

That Bertrand Piccard was attracted to the challenge was quite natural. An experienced aviator who has flown everything from ultralights to high-performance aerobatic aircraft, he is the grandson of Auguste Piccard, who invented the pressurized balloon gondola, and flew a balloon into the stratosphere for the first time on May 27, 1931.

DIMENSIONS: ——————

WIDTH	5.20 M (15 FT.)
LENGTH	2.3 M (7 FT.)
HEIGHT	2.5 M (4 FT., 6 IN.)
WEIGHT	2,000 KG (4,409 LB.)

1. The *Breitling Orbiter 3* balloon in flight over the Alps, not long after takeoff from Château d'Oex, Switzerland, on March 1, 1999. **2.** Pilots Bertrand Piccard (left) and Brian Jones (right) pose waving atop the gondola of their balloon *Breitling Orbiter 3* on display in the Milestones of Flight gallery at the National Air and Space Museum. **3.** The official insignia for the *Breitling Orbiter 3* round-the-world balloon flight, March 1999.

The lessons learned on the failed *Breitling Orbiter 1* and *2* flights enabled Piccard and his team to develop technical systems and a basic strategy for the *Breitling Orbiter 3* project.

Designed and built by Cameron Balloons, of Bristol, England, *Breitling Orbiter 3* stood 55 m (180 ft.) tall when fully inflated. Like most of the balloons entered in the around-the-world competition, it combined the advantages of hot-air and helium technologies. The envelope was constructed of a nylon fabric welded to a helium-tight membrane covered with an outer protective skin coated with aluminum on both sides to provide improved thermal control. The shape and special features of the envelope ensured maximum temperature stability in order to maintain a constant altitude, conserve helium, and reduce propane consumption.

The propane gas that fueled the six burners of *Breitling Orbiter 3* was contained in twenty-eight titanium cylinders mounted in two rows along the sides of the gondola. Concerned about fuel consumption, the team added four additional propane containers prior to takeoff. It was a wise decision. When the balloon landed, there was less than a quarter of a tank of fuel remaining.

The *Breitling Orbiter 3* gondola was constructed of a weave of Kevlar and carbon fiber material. After takeoff the cabin was sealed at 1,829 m (6,000 ft.) to trap the air within it. During the flight, the cabin atmosphere was supplemented by oxygen, and the carbon dioxide was removed by lithium hydroxide filters. Cabin pressure was maintained at around 3.5 psi by adding oxygen and nitrogen to the cabin air as necessary. Solar panels suspended beneath the gondola recharged the onboard lead-acid batteries that provided electrical power. Satellite-based systems enabled the crew to communicate and navigate.

Piccard and Jones guided *Breitling Orbiter 3* up and away from the Swiss Alpine village of Château d'Oex at 8:05 a.m. GMT on March 1, 1999. They landed in the Egyptian desert 19 days, 21 hours, and 55 minutes later, having traveled a distance of 40,814 km (25,361 mi.). The balloon had climbed to altitudes of up to 11,373 m (37,313 ft.) and achieved a maximum speed of 161 knots per hour.

Thanks to the then National Air and Space Museum director, the late Adm. Donald Engen, who had superb negotiating skills and wide contacts in aviation and business, *Breitling Orbiter 3* now rests in a place of honor in the Milestones of Flight gallery.

1

ALTHOUGH THE CAUDRON G.4 has great significance as an early light bomber, its most influential role in World War I was that of reconnaissance.

Brothers Gaston and René Caudron were among the earliest aircraft manufacturers in France. After building and testing a few original designs in 1909 and early in 1910, they established a flight-training school at Crotoy and an aircraft factory at Rue in 1910. The first factory-produced Caudron was the type A4, a 35-hp Anzani-powered tractor biplane in which the pilot sat completely exposed behind the rear spar of the lower wing.

The next major design, the type B, was the first Caudron to feature the abbreviated fuselage-pilot nacelle. It was powered by a 70-hp Gnôme or 60-hp Anzani engine mounted in the front of the nacelle with the pilot immediately behind. Although a tractor, the tail unit of the type B was supported by booms extending from the trailing edge of the wings, an arrangement more commonly featured on pusher aircraft. Lateral control was accomplished with wing warping. The type B established the basic configuration of Caudron design through the G.4 model.

The first of the well-known Caudron G series aircraft appeared in 1912. Initially designed as a trainer, the type G was developed into the G.2 by the outbreak of the First World War and saw limited military service in 1914 as single- and two-seat versions. By that time the Caudron factory had been relocated to Lyon, where an improved version, designated the G.3, was being produced in significant numbers. Soon a second factory was opened at Issy-les-Moulineaux near Paris to meet military demand. The G.3 was primarily a two-seat aircraft, but a few were converted to single-seat. They were powered variously by 80-hp Le Rhône or Gnôme rotary engines or a 90-hp Anzani radial. A total of 2,450 G.3s were

1. NASM's restored Caudron G.4 at the Paul Garber Preservation, Restoration, and Storage Facility.
2. A Caudron G.4 in very low-level banking flight over the Caudron factory at Issy-les-Moulineaux, Paris, France, circa 1915.
3. Two members of the French Escadrille C.11 stand in front of a Caudron G.4 on February 26, 1917. 4. Note the sparse forward cockpit for the gunner-bombardier.

DIMENSIONS: ――――

WINGSPAN	17.2 M (56 FT., 5 IN.)
LENGTH	7.2 M (23 FT., 8 IN.)
HEIGHT	2.6 M (8 FT., 6 IN.)
WEIGHT	733 KG (1,616 LB.)

built, including a small number built under license in Britain and Italy.

The Caudron G.4 was a larger, twin-engine version of the G.3, powered by two 80-hp Le Rhônes or 100-hp Anzanis.

The twin-engine configuration increased the range of the Caudron and provided a location for a forward-firing machine gun. To protect against attacks from behind, some G.4s were fitted with an additional gun mounted on the top of the upper wing and pointed rearward. A number of G.4s had a second gun mounted immediately in front of the pilot on the deck of the nacelle. Most often the pilot and observer simply carried handheld weapons to respond to attacks from the rear. Some G.4s carried a camera for high-altitude reconnaissance.

The prototype G.4 first flew in March 1915, and 1,358 were built in three major versions: the Caudron G.4A2 for reconnaissance, which had a wireless set for artillery-spotting mission; the G.4B2, which could carry a 100-kg (220-lb.) load of bombs; and the G.4E2, which had dual controls for training. A special armored version of the G.4, designated the G.4IB, was deployed to the top French units. The B represented *Blindage*, the French word for "armor." The Caudron G.4 also sometimes served as a long-range escort to other bomber aircraft.

Extensively used as a bomber during the first half of 1916, the Caudron's relatively slow speed and inability to defend itself from the rear made it increasingly vulnerable to fighter attack as German air defense improved. Caudrons continued to be widely used as reconnaissance aircraft well into 1917. By early 1918 virtually all Caudron aircraft still in use were relegated to training duties.

In addition to the French, Caudrons were used extensively by British and Italian units, and a few were used by the Russians and the Belgians. Ten Caudron G.4s were sold to the United States. Used exclusively as trainers, none of these Caudrons saw operational service with American units.

The Caudron G.4 in the National Air and Space Museum is among the oldest surviving bomber aircraft in the world, and it is the only multiengine airplane from this period anywhere. It was purchased by the US government in 1917 and transferred to the Smithsonian in 1918.

ON APRIL 17, 1964, Geraldine Mock in her Cessna 180 *Spirit of Columbus* made history as the first woman to pilot an aircraft around the world.

Mock learned to fly in 1962 and immediately began planning for her epic trip. She quickly acquired her instrument rating and, by the time she had flown 750 hours, was ready for her around-the-world attempt.

The Cessna 180 series and its descendants are another great success story for Cessna Aircraft Corporation of Wichita, Kansas, and general aviation. Borrowing some features, such as flaps, from the army's L-19 Bird Dog liaison airplane and evolving directly from the postwar Model 170B four-seat design, the 180, introduced in 1952, became a rugged four-place, high-wing plane with a top speed of 266 km/h (165 mph). Variations on the basic airframe and 225- or 230-hp engines made it a popular bush-type aircraft in most of the undeveloped parts of the world as well as in the United States. The Model 180 was also successfully marketed as a sophisticated business and personal aircraft.

In 1956, by adding a tricycle landing gear, the 180 became the Model 182 Skylane, which, with cockpit and exterior improvements, enjoyed a long initial production run as a high-performance workhorse in the personal, business, and utility categories. The Cessna 185 Skywagon was introduced in 1960 with a more powerful engine.

Cessna built more than 6,000 of the Model 180 series aircraft and 22,000 Model 182s through 1986. Cessna halted production after a product liability crisis and other economic issues threatened the future of small light planes. However, with the passage of the General Aviation Revitalization Act in 1994, the company began producing the Model

DIMENSIONS:

WINGSPAN	10.97 M (36 FT.)
LENGTH	7.98 M (26 FT., 2 IN.)
HEIGHT	2.36 M (7 FT., 9 IN.)
WEIGHT	1,157 KG (2,550 LB.)

1. The Cessna 180 *Spirit of Columbus* hanging on display at the National Air and Space Museum's former General Aviation Gallery. **2.** Geraldine L. "Jerrie" Mock, standing beside her Cessna 180 *Spirit of Columbus* at Port Columbus Airport, Columbus, Ohio, March 19, 1964, immediately before she took off on her record-setting solo flight around the world. **3.** The Cessna 180 *Spirit of Columbus,* taxing into position in front of King Fahd Dhahran Air Terminal, Dhahran, Saudi Arabi, April 1964.

182 (and other single-engine models) again in July 1996 at a new factory in Independence, Kansas.

For Mock's world flight, additional fuel tanks were custom-made to fit inside the cabin, which provided a total of 674 l (178 gal.) of gasoline—enough to fly for 24 hours and over 4,184 km (2,600 mi.).

On March 19, 1964, at 9:31 a.m., Mock departed Columbus, Ohio, and headed southeast toward Bermuda, her first stop on her history-making solo flight. From there she flew on to Morocco, Algeria, Libya, Egypt, Saudi Arabia, and Pakistan. From there she continued on to India, Thailand, the Philippines, Guam, Wake Island, and Hawaii. The longest leg of the flight was the 3,862 km (2,400 mi.) between Hawaii and Oakland, California, which she accomplished without incident in 18 hours.

From Oakland the flight was uneventful as Mock flew on to Tucson, El Paso, Bowling Green, Kentucky, and finally back to Columbus, arriving 29 days, 11 hours, and 59 minutes later on April 17, 1964, after flying 37,181 km (23,103 mi.). The flight was monitored by the National Aeronautic Association and the Fédération Aéronautique Internationale, which certified it as an around-the-world speed record for aircraft weighing less than 1,750 kg (3,858 lbs.). President Lyndon B. Johnson awarded Mock the Federal Aviation Administration's Exceptional Service Decoration.

The record-setting aircraft was given to the National Air and Space Museum in 1975. It was displayed in the General Aviation gallery until 1984 and will eventually be housed at the Steven F. Udvar-Hazy Center.

FOR TWENTY-SEVEN YEARS, the Concorde carried world travelers across the Atlantic Ocean in great comfort at twice the speed of sound. High development and operating costs, however, prevented the Concorde from becoming a practical means of supersonic flight for the public.

Enterprising engineers in Great Britain and France were independently designing airplanes for a supersonic transport (SST). In November 1962 the two nations agreed to pool their resources and share the risks in building this new aircraft. They also hoped to highlight Europe's growing economic unity and supplant the United States as the leader in commercial aviation. The name Concorde reflected the shared hopes of each nation for success through cooperation.

Designers at the British Aircraft Corporation and Sud Aviation, later reorganized as Aérospatiale, settled on a slim, graceful form that featured an ogival delta wing that possessed excellent low-speed and high-speed handling characteristics. Power was to be provided by four massive Olympus turbofan engines built by Rolls-Royce and SNECMA.

Realizing that this first-generation SST would cater to the wealthier passenger, Concorde's designers created an aircraft that carried only one hundred seats in tight four-across rows.

Despite mounting costs that constantly threatened the program, construction continued with exactly 50 percent of each aircraft being built in each country. The first of four Concorde prototypes was flown by famed French test pilot Andre Turcot on March 2,

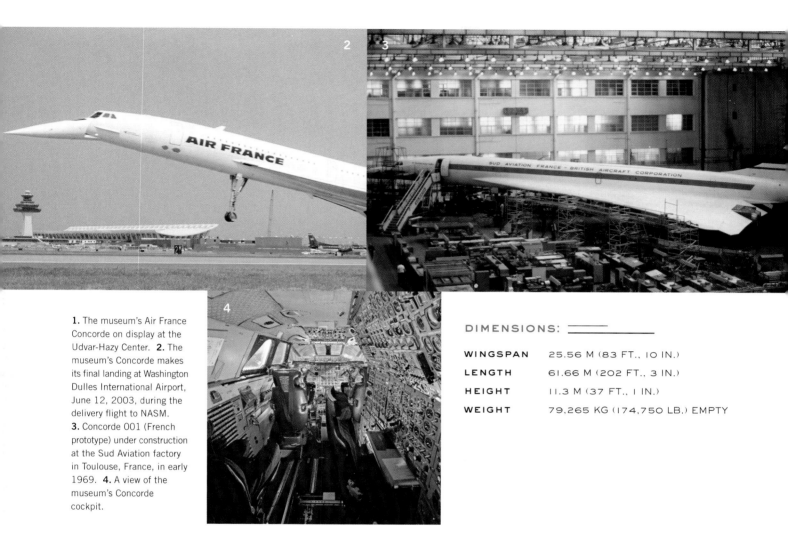

1. The museum's Air France Concorde on display at the Udvar-Hazy Center. 2. The museum's Concorde makes its final landing at Washington Dulles International Airport, June 12, 2003, during the delivery flight to NASM. 3. Concorde 001 (French prototype) under construction at the Sud Aviation factory in Toulouse, France, in early 1969. 4. A view of the museum's Concorde cockpit.

DIMENSIONS:

WINGSPAN	25.56 M (83 FT., 10 IN.)
LENGTH	61.66 M (202 FT., 3 IN.)
HEIGHT	11.3 M (37 FT., 1 IN.)
WEIGHT	79,265 KG (174,750 LB.) EMPTY

1969. After thorough testing, the first production Concordes were ready for service in 1976.

All was not rosy, however. In the United States the federal government refused to back the project, citing environmental problems like noise, the sonic boom, and engine emissions that were thought to harm the upper atmosphere. Anti-SST politics delayed the granting of landing rights, particularly into New York City, causing further delays.

Despite initial enthusiasm, no airlines placed orders after calculating the expense of operating the Concorde. Only after negotiating to purchase them from their governments at virtually no cost did Air France and British Airways buy the sixteen production aircraft. In January 1976, Concorde service began and by November they were flying to the United States.

A technological masterpiece, these planes smoothly transitioned to supersonic flight with no discernible disturbance to the passenger, cruising at between 16,764 and 18,288 m (55,000 and 60,000 ft.)—so high that passengers could actually see the curvature of the Earth. The Concorde was so fast that, despite an outside temperature of less than −56°C (−69°F), its aluminum skin heated to over 120°C (248°F) and actually expanded 8 inches in length. Transatlantic flight time was cut in half with the average flight taking less than four hours.

Eventually the harsh reality of the economic marketplace forced Air France and British Airways to cut back their already limited service. Routes from London and Paris to Washington, Rio de Janeiro, Caracas, Miami, Singapore, and other locations were cut, leaving only the transatlantic service to New York. Most of these flights flew half full. With the average round-trip ticket costing more than $12,000, few could afford it. In 2000 the only fatal Concorde accident temporarily grounded the fleet. Concorde service ended on May 31, 2003.

Air France gave Concorde F-BVFA to the National Air and Space Museum in June 2003.

ON MAY 29, 1909, the G. H. Curtiss Manufacturing Company delivered their first airplane to the New York Aeronautical Society for $5,000. Called the *Golden Flyer* because of its golden-yellow silk covering and the orange shellac coating, this was the first nonmilitary sale of an airplane in the United States.

The *Golden Flyer* was a single-seat pusher aircraft with the engine and propeller behind the pilot with single-surface wings, a biplane forward elevator on long forward booms, and a fixed horizontal stabilizer and rudder on long rear booms. The airplane sat on a three-wheel landing gear with the nose wheel fairly far forward, a feature intended to prevent nose-overs.

Directional control of the airplane was accomplished by turning a steering wheel on the control column left or right, climb and descent were controlled by fore and aft movement of the column, and roll was achieved by leaning left or right against a shoulder yoke that actuated the ailerons. The airplane was powered by a 25-hp, four-cylinder Curtiss engine, driving a single six-foot laminated wooden propeller. Before delivering the *Golden Flyer,* Glenn Curtiss won the Scientific American Trophy with it for a second time in 1909. Curtiss was selected by the Aero Club of America to be the sole American in the first international aviation meet in Reims, France, in August 1909. For this event, he designed and built a new airplane that came to be known as the *Reims Racer.*

Similar to the *Golden Flyer,* the *Reims Racer* had a shortened wingspan and was covered with gray silk fabric. It was powered by a 50-hp Curtiss V8 engine. Curtiss won the coveted Gordon Bennett Aviation Cup at Reims with an average speed of 77 km/h (47.6 mph). With prize money of $7,600, he went on to Brescia, Italy, where he won the grand prize and altitude prize, adding another $7,000 to his winnings. Curtiss returned to the United States an international hero.

DIMENSIONS:

WINGSPAN	11.6 M (38 FT., 1 IN.)
LENGTH	7.8 M (25 FT., 6 IN.)
HEIGHT	2.7 M (9 FT.)
WEIGHT	632 KG (1,390 LB.)

1. The museum's Curtiss D-III Headless Pusher at the Paul E. Garber Facility, following its completed restoration on October 30, 1979. **2.** A Curtiss Pusher piloted by Lincoln Beachey racing Barney Oldfield in his automobile in Columbus, Ohio, 1914. **3.** Lincoln Beachey seated at the controls of his specially modified Curtiss D "Beachey Special." **4.** Eugene B. Ely landing his Curtiss D biplane on a platform constructed on the USS *Pennsylvania* anchored in San Francisco Bay, January 18, 1911. This was the first time an airplane landed on a ship.

In 1911, Curtiss pilot Eugene Ely had created a sensation when he made a successful takeoff from the USS *Birmingham* on November 14, 1910. Topping that, on January 18, 1911, Ely landed a Curtiss pusher on a platform built on the after deck of the cruiser USS *Pennsylvania*. The US Navy purchased three Curtiss airplanes.

Curtiss continued the evolution of the pusher design with the development of the D-II (the *Golden Flyer* was considered the Model D), which relocated the ailerons from the front interplane struts to the rear ones, improving the efficiency of the wings and the ailerons.

The Curtiss D-III incorporated covering on both the top and bottom surfaces of the wings, enclosing the ribs and spars and adding 4.8 to 9.7 km/h (3 to 6 mph) in speed. The forward elevator was moved back slightly and placed almost directly above the front wheel, and elevators were added to the rear in place of the fixed horizontal stabilizer.

The Curtiss D-III Headless Pusher—so called because the forward elevator had been removed—was the result of a propitious accident incurred by noted pilot Lincoln Beachey. While landing from an exhibition flight, Beachey hit a fence and destroyed the front elevator. Beachey flew on without the front elevator control and found to his pleasant surprise that the aircraft performed better than before. As it turned out, navy pilots had independently realized that stability was enhanced without the forward elevator and they removed them from their airplanes. Curtiss concurred with the results and began producing the 1912 Model D Headless Pusher as a new offering.

Curtiss developed the world's first successful seaplane pusher, or hydroaeroplane as he called it, in 1911 when he fitted a standard pusher with pontoons. Naval seaplanes were fitted with a single pontoon and wingtip floats, and civilian seaplanes had twin pontoons. Curtiss was the first recipient of the Collier Trophy in 1912 for his seaplane.

The Curtiss D-IV variant intended for the military market appeared in 1911. Essentially the same as the D-III, except for increased wingspan and the addition of a passenger seat behind that of the pilot, the D-IV was designed to be quickly dismantled for easy transport.

In 1919 a replica of a 1912-style Curtiss Headless Pusher was constructed at a Curtiss research facility in Garden City, New York, under the personal direction of Glenn Curtiss. The replica was donated to the Smithsonian in 1925.

ONE OF THE most famous of all aircraft is the Curtiss JN-4D, popularly known as the Jenny. More than 90 percent of American pilots trained during World War I received their primary instruction on the Curtiss Jenny. And, thanks to Jenny, the era of barnstorming brought airplanes to many in the 1920s.

The Curtiss JN series began in 1914 as a hybrid, incorporating the best features of the Model N, designed in America by Glenn Curtiss, and the Model J designed in England by a young engineer named B. Douglas Thomas. Thomas's Model J and Curtiss's Model N were both entered in the US Army evaluation trials in September 1914, from which Curtiss received an order for eight modified Model Js. Curtiss added the letter N to the designation because the modified Model J incorporated significant features of the Model N. The designation officially became the JN-2.

The First Aero Squadron of the US Air Service, Signal Corps, deployed JN-2s to Mexico in 1916 where they performed tactical operations for the campaign against Pancho Villa. To improve performance, Curtiss changed the wings and ailerons, replaced the yoke control system with a wheel, and added a foot bar to control the rudder. The improved design was designated the JN-3. This model caught the interest of the British Royal Naval Air Service. With a few further revisions, JN-4s were acquired by the British as trainers, by the US Air Service, by the Curtiss flying schools, and by private owners.

DIMENSIONS: ‗‗‗‗

WINGSPAN	13.3 M (43 FT., 7.125 IN.)
LENGTH	8.3 M (27 FT., 4 IN.)
HEIGHT	3 M (9 FT., 10.625 IN.)
WEIGHT	718 KG (1,580 LB.)

1. The museum's US Army Air Service Curtiss JN-4D Jenny on display at the Paul E. Garber Facility. **2.** Museum Specialist Lynn Wilson (left) and Stanley Potter paint the insignia of 46th Squadron, Aviation Section, Signal Corps. **3.** Two Curtiss JN-4B Jenny biplanes in flight.

The JN-4A had larger and redesigned tail surfaces, a revised fuselage, increased dihedral, and ailerons on both wings to improve lateral control. Canadian-built Jennys (the JN-4 [Can], or the Canuck) incorporated a control stick instead of a wheel, a revised tail, and strut-connected ailerons on both wings. The JN-4D, the definitive design of the JN series, was introduced in June 1917. Principal design changes for this model included a control stick, ailerons only on the upper wing, and curved cutouts on the inner trailing edges of all four wing panels.

The need for an advanced trainer led to the JN-4H with a 150-hp Wright-Martin–built, Hispano-Suiza engine. The JN-4HT was a dual-control trainer, the JN-4HB was a bomber trainer fitted with bomb racks, and the JN-4HG was a single-control gunnery trainer. Minor improvements in the JN-4H series led to the JN-6H with a reversion to strut-connected ailerons on both wings.

The final iteration of the JN series was the JNS, or Standard Jenny. This designation, appearing in 1923, was used to identify obsolescent JN-4H and JN-6H models that were rebuilt and modifications to ailerons on the top wing.

During World War I, Curtiss and six other American companies delivered 6,070 JN series aircraft to the US Air Service.

At the end of the war, the US government began to sell their surplus Jennys.

By these means, the Jenny became the principal aircraft flown by barnstormers in the 1920s. Americans, particularly in rural areas, thrilled to the antics of these pilots performing in the aerial circuses that toured the country. For many, the Jenny was the first airplane they had seen close-up, and those with a few dollars and their fear in check typically would make their first flight in a Jenny. The slow-flying Jenny was perfect for wing walkers, who clung to the Jenny's maze of struts, and performed death-defying stunts. In 1926, newly instituted federal airworthiness requirements for both airplanes and pilots brought the era of the Jenny and barnstorming to a close.

In November 1918 the Smithsonian acquired a JN-4D Jenny from the US Army Air Service, which had used it briefly as a trainer.

WHETHER IT WAS called the Tomahawk, Warhawk, or Kittyhawk, the Curtiss P-40, in all of its many variations, was a rugged and effective fighter available in large numbers early in World War II when America and her allies urgently needed them.

Engineer Donovan R. Berlin laid the foundation for the P-40 in 1935 when he designed the agile but lightly armed P-36 fighter. It was soon obvious, however, that the P-36 was no match for newer European fighters in high-altitude combat.

Modifications were made, including the installation of a more powerful Allison V-1710 engine in place of the Pratt & Whitney R-1830, and the US Army Air Corps ordered 540. It was given an armament package consisting of two .50-caliber machine guns in the fuselage and four .30-caliber machine guns in the wings, and was designated the P-40.

After production began in 1940, France ordered 140 P-40s, but the British took delivery of these airplanes when Paris surrendered to Germany. The British named the aircraft Tomahawks but found they performed poorly in high-altitude combat over northern Europe and relegated them to low-altitude operations in North Africa. The Russians acquired more than 2,000 P-40s.

When the United States declared war in 1941, P-40s equipped most of the Army Air Corps' frontline fighter units. This fighter eventually saw combat in almost every theater of operations, but it was most effective in the China-Burma-India (CBI) Theater. Of all the CBI groups, it was the P-40s of a Nationalist Chinese Air Force unit that remain to this day synonymous with the P-40: the American Volunteer Group (AVG) and its successor, the Flying Tigers.

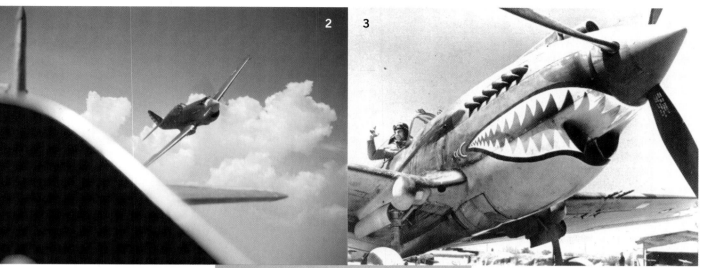

1. The museum's Curtiss P-40E Warhawk *Lope's Hope* hanging on display at the Udvar-Hazy Center. 2. A close-up view of a Headquarters Squadron 8th Pursuit Group Curtiss P-40E Warhawk. 3. An American Volunteer Group (AVG) Curtiss P-40E Warhawk on the ground, showing off its shark mouth markings. 4. A formation of three Curtiss P-40E Warhawks in flight over in the vicinity of Randolph Field, Texas.

DIMENSIONS:	
WINGSPAN	11.37 M (37 FT., 4 IN.)
LENGTH	9.49 M (31 FT., 2 IN.)
HEIGHT	3.23 M (10 FT., 7 IN.)
WEIGHT	2,880 KG (6,350 LB.)

Organized by the US Army Air Corps' Capt. Claire Chennault, the Flying Tigers—the name derived from the ferocious fangs and teeth painted on the nose of AVG P-40s at either side of the distinctive, large radiator air intake—were the first real opposition the Japanese military encountered. In less than seven months of action, AVG pilots destroyed about 115 Japanese aircraft and lost only 11 planes in air-to-air combat.

With wartime experience, Curtiss made many modifications, including adding armor plate, improved self-sealing fuel tanks, and a more powerful engine. They changed the cockpit to improve visibility and changed the armament package to six wing-mounted .50-caliber machine guns. The P-40E Kittyhawk was the first model with this configuration, and it entered service in time to serve in the AVG.

In addition to the AAF, England, France, China, Russia, Australia, New Zealand, Canada, South Africa, and Turkey flew P-40s in the war.

The Smithsonian's P-40E did not serve in the US military. Built in 1941, it served in No. 111 Squadron, Royal Canadian Air Force. The US Air Force restored it in 1975 and painted it to the colors of the 75th Fighter Squadron, 23rd Fighter Group, 14th Air Force before donating it to the National Air and Space Museum.

CURTISS R3C-2

1

EARLY IN THE development of aviation a spirit of sporting and competition became a major aspect of its ever-growing appeal. Air races began to enjoy a worldwide popularity, and two of the most coveted prizes were the Pulitzer Trophy and the Schneider Cup.

In 1925 the US Army and Navy ordered from the Curtiss Aeroplane and Motor Company airplanes of the same design but with individual variations. These airplanes ran away with first place in both the Pulitzer Trophy Race and in the Schneider Cup Race in that same year.

Lt. Cyrus Bettis piloted the Curtiss R3C-1 that won the Pulitzer Trophy Race on October 12, 1925, at a speed of 401 km/h (248.9 mph). On October 25, Lt. James H. "Jimmy" Doolittle won the Schneider Cup Race held at Bay Shore Park in Baltimore with a speed of 374 km/h (232.57 mph) in a R3C-2, the seaplane version of the R3C-1. On the next day, Doolittle flew the R3C-2 over a straight course at a world-record speed of 395 km/h (245.7 mph).

The next year, this same R3C-2 piloted by Lt. Christian F. Schilt and powered by an improved engine, finished in second place with an average speed of 372 km/h (231.4 mph).

The R3C-2 was a single-seat, single-bay, wire-braced biplane. The wings were covered with two-ply spruce planking .094-inch thick, forming a box structure that required no internal bracing. Among the interesting features were the low-drag wing radiators made of corrugated, .004-inch-thick brass sheeting covering much of the surface of both upper and lower wings with the corrugations running chordwise. The upper wing was flush with the top of the fuselage, permitting the pilot to see over the wing. All ribs were of spruce, and the

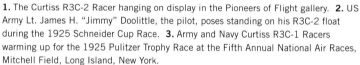

DIMENSIONS: ═════

WINGSPAN	6.71 M (22 FT.), UPPER;
	6.1 M (20 FT.) LOWER
LENGTH	6.01 M (19 FT., 8.5 IN.)
HEIGHT	2.46 M (8 FT., 1 IN.)
WEIGHT	975 KG (2,150 LB.)

1. The Curtiss R3C-2 Racer hanging on display in the Pioneers of Flight gallery. **2.** US Army Lt. James H. "Jimmy" Doolittle, the pilot, poses standing on his R3C-2 float during the 1925 Schneider Cup Race. **3.** Army and Navy Curtiss R3C-1 Racers warming up for the 1925 Pulitzer Trophy Race at the Fifth Annual National Air Races, Mitchell Field, Long Island, New York.

ailerons, made of metal, were fabric-covered. The cantilever vertical fin and horizontal stabilizer were wooden.

An ingenious streamlined monocoque structure, the fuselage consisted of a shell of two layers of spruce over which fabric was doped for added strength and protection. This shell was formed over seven birch plywood bulkheads that were connected by four ash longerons, making the structure rigid.

The unbalanced movable controls were metal. Only necessary navigation and engine instruments were installed. They consisted of gauges for water temperature, oil temperature, oil pressure, and fuel quantity, as well as a

tachometer and an airspeed indicator. The fixed landing gear in the R3C-1 was a tripod configuration. A laminated hickory tail skid was added to protect the rudder.

The R3C-1 carried only 102 l (27 gal.) of fuel, which gave 48 minutes flying time at full throttle. In the R3C-2, the fuel capacity, achieved by installing fuel tanks in the floats, was 227 l (60 gal.), enough for 1.3 hours at full throttle.

The R3C-2 in the National Air and Space Museum is painted in the colors of the one flown by Doolittle in the 1925 Schneider Trophy Race. It was on loan for several years to the Air Force Museum, where they restored it. It now hangs in the Pioneers of Flight gallery.

THE DASSAULT FALCON 20C is a French executive jet aircraft originally developed as the commercial version of the famous Mystère fighter aircraft that is part of an elite group of such aircraft as the Lear Jet (see page 100), the Hawker-Siddeley (now British Aerospace) 125, and the North American Sabreliner.

In the United States, the Falcon 20C was marketed as the Fan Jet Falcon by Pan American Airways through its subsidiary, the Falcon Jet Corporation, which was established in 1972 for the specific purpose of selling this aircraft to the US market.

The first Dassault Falcon made its maiden flight in May 1963. It is a well-proportioned, all-metal, low-wing monoplane with full cantilever wing and tail surfaces, pressurized fuselage, and retractable tricycle dual-wheel landing gear. It is powered by two aft-mounted General Electric CF-700-2D turbofan engines.

The Falcon 20C in the National Air and Space Museum was modified by Federal Express for cargo use. The changes to the 20 series included the installation of a cargo door measuring 1.4 by 1.9 m (55 by 74.5 in.) on the left side of the forward fuselage. Control of

DIMENSIONS: ═══════

WINGSPAN	16.3 M (53 FT., 6 IN.)
LENGTH	17.2 M (56 FT., 4 IN.)
HEIGHT	5.7 M (17 FT., 7 IN.)
WEIGHT	7,230 KG (15,940 LB.)

3

1. NASM's Federal Express Dassault Mystère/Falcon 20 *Wendy* on display at the Udvar-Hazy Center. The tail of Air France Concorde is visible in background. 2. A Federal Express Dassault Mystère/Falcon 20 *Ann-Marie* loading and unloading cargo from a Federal Express van. 3. The museum's Federal Express Dassault Falcon 20 *Wendy* making a landing.

the door is achieved by a closed-circuit electrohydraulic system that utilizes the aircraft batteries to operate an electric motor and hydraulic pump. The floor was strengthened to accept the extra weight, and the passenger windows were plugged.

Federal Express service represented a milestone of air transport in the United States. FedEx founder Fred Smith created this new category of airline service when he launched his package distribution system, which is based on the concept of a centralized clearinghouse (and is located in Memphis, Tennessee). The first two Falcons were delivered in June 1972 and cost $1.2 million each. The service was an immediate success, and within a few months more Falcons were ordered.

The tail number, N8FE, does not mean that it was the eighth in the line. In fact, it was the first one delivered, but Fred Smith felt that no harm would be done if the public assumed that Federal Express already had a fleet of eight airplanes.

Because of Federal Express's success, the Falcon 20Cs were replaced by a fleet of larger Boeing 727-100Fs and eventually McDonnell Douglas DC-10Cs—whose cargo holds were big enough to carry several Falcons each.

The Falcon 20C that Federal Express donated to the National Air and Space Museum in 1983 carried the company's first package. The plane was named *Wendy* after Fred Smith's daughter. The Falcon reminds us that air transport is as dynamic today as it ever was.

1

THE DE HAVILLAND Chipmunk was originally a post–World War II primary trainer, but veteran aerobatic pilot Art Scholl used to thrill audiences with skill and showmanship in his *Pennzoil Special*.

The DHC-1A was designed, built, and flown by de Havilland's Canada subsidiary, hence the Canadian woodsy sounding name of Chipmunk. The prototype first flew on May 22, 1946, in Toronto. De Havilland of Canada produced 158 Chipmunks, and de Havilland of England produced 740 airplanes for training for the Royal Air Force during the late 1940s and into the 1950s. In 1952 the Duke of Edinburgh took his initial flight training in a Chipmunk. It was also used for light communications flights in Germany and for internal security duties on the island of Cyprus.

The Chipmunk was an all-metal, low-wing, tandem (two-place), single-engine airplane with a conventional tail-wheel landing gear. It had fabric-covered control surfaces and a clear plastic canopy covering the pilot and passenger positions. The production versions of the airplane were powered by a 145-hp in-line de Havilland Gipsy Major "8" engine.

Art Scholl, an aerobatic and movie pilot, purchased two Canadian-built Chipmunks from the surplus market in the late 1950s and early 1960s. In 1968 he began extensive modifications to one that resulted in an almost completely new aircraft.

He covered over one cockpit and installed a fuel-injected 260-hp Lycoming GO-435 flat-opposed six-cylinder engine. He removed 25 cm (20 in.) from each wingtip and changed the airfoil section of the tip area. The reduction in span led to the need to lengthen

1. The museum's de Havilland DHC-1A Chipmunk hanging on display at the Udvar-Hazy Center, 2004. **2.** An aerial view of the museum's de Havilland DHC-1A Chipmunk in flight over snow-covered mountainous area, piloted by aerobatic pilot Art Scholl. **3.** Judy Scholl working on the engine of the Chipmunk with the aircraft's insignia clearly visible. **4.** The pilot of the museum's Chipmunk, Art Scholl, with his dog Aileron perched on his shoulder.

DIMENSIONS: ————

WINGSPAN	7.6 M (24 FT., 9 IN.)
LENGTH	6 M (20 FT.)
HEIGHT	1.8 M (6 FT.)
WEIGHT	2,530 KG (1,150 LB.)

the ailerons inboard to retain control effectiveness. This in turn reduced the flaps to where they became somewhat ineffective, and, since the flaps really were not required for his aerobatic routines, he removed them as a weight-saving measure. These modifications improved the low-speed tip stall characteristics and improved roll performance.

Other idiosyncrasies were the pitot static tube that was fashioned from a golf club shaft and a 7.6-cm (3-in.) extension added to the cockpit control stick to ease the control loads during the more severe aerobatic routines. Scholl also installed rearview mirrors on both sides of the cowling just forward of the windscreen. He installed three smoke generators with red, white, and blue smoke for his show routines, which included the Lomcevak tumbling/tailslide maneuver.

A PhD in aeronautics, Scholl designed most of these modifications himself. He held all pilot ratings, and he was

a licensed aircraft and powerplant mechanic and an authorized FAA inspector. He was also a three-time member of the US Aerobatic Team, an air racer (placing several times at the National Air Races at Reno), an air show pilot, and a fixed-base operator at a school of international aerobatics.

In 1959, Scholl began working for legendary Hollywood pilots Frank Tallman and Paul Mantz at Tallmantz Aviation and then later formed his own movie production company, producing and performing aerial photography and stunts for many movies and television shows. At air shows, Scholl often flew with his dog, Aileron, who rode the wing as Scholl taxied on the runway or sat on his shoulder in the aircraft.

Art Scholl was killed in 1985 while filming *Top Gun.* Art Scholl's estate donated his *Pennzoil Special* to the Smithsonian on August 18, 1987.

ON THE MORNING of November 20, 1953, A. Scott Crossfield became the first pilot to fly at twice the speed of sound in the experimental air-launched rocket-propelled Douglas D-558-2 #2 Skyrocket.

Shortly after reaching Mach 2.005, or 2,078 km/h (1,291 mph) while in a shallow dive at an altitude of 18,897.6 m (62,000 ft.), the plane's XLR-8 rocket engine exhausted its fuel supply and shut down. Crossfield glided earthward to a smooth dead-stick landing on Muroc Dry Lake, at Edwards Air Force Base, California.

The D-558-2 #2 was just one of six different D-558 research airplanes ordered by the US Navy from the Douglas Aircraft Company for obtaining information about aerodynamics at transonic and supersonic speeds. Analysis of data captured from the Germans on their wartime swept-wing research, combined with studies by American scientist Robert T. Jones, caused Douglas and the navy to concentrate on swept-wing vehicles powered by both turbojet and rocket engines. The first three aircraft, each powered by a single General Electric TG-180 turbojet, became known as the D-558-1 Skystreak series. The last three, powered initially by a Westinghouse J-34 turbojet for low-speed flight plus a Reaction Motors XLR-8 rocket engine for high-speed research, became known as the D-558-2 Skyrocket series.

Because of its engine type and airframe design, the D-558-1 was limited to approximately Mach 1. The more powerful D-558-2, with its 6,000-pound-thrust rocket engine fueled with liquid oxygen and diluted ethyl alcohol, could easily exceed Mach 1. For safety reasons, Douglas modified the D-558-2 #2 and #3 for air launching from the bomb

DIMENSIONS: ──────

WINGSPAN	7.62 M (25 FT.)
LENGTH	12.80 M (42 FT.)
HEIGHT	3.86 M (12 FT., 8 IN.)
WEIGHT	4,673 KG (9,421 LB.)

1. The museum's NACA Douglas D-558-2 Skyrocket #2 hanging on display on second-floor level, at the National Air and Space Museum. **2.** NASM's Douglas D-558-2 Skyrocket moments after being released from its US Navy Boeing P2B-1S (B-29) Superfortress mother ship over Edwards AFB, Muroc, California, on April 18, 1951. **3.** An unidentified NACA researcher makes adjustments to Douglas D-558-2 Skyrocket scale model inside test cell of the 4-by-4-foot supersonic pressure tunnel at Langley Aeronautical Laboratory, Virginia. **4.** A view of Douglas D-558-2 Skyrocket in flight over desert near Edwards AFB, Muroc, California, in November 1949.

bay of a converted Boeing P2B-1S (Navy B-29) Super-fortress (see page 28). At the same time, Douglas modified the D-558-2 #2 to all-rocket propulsion, utilizing the space formerly taken up by its turbojet engine for additional rocket fuel. The modified D-558-2 #2 reached Mach 2 during a special high-speed flight by the National Advisory Committee for Aeronautics (NACA).

On August 20, 1947, Navy Cmdr. Turner F. Caldwell set a new world air-speed record of 1,031.05 km/h (640.663 mph) while flying the D-558-1 #1. Five days later, on August 20, Marine Corps Maj. Marion Carl flew the second D-558-1 a record-breaking 1,047.35 km/h (650.796 mph). NACA utilized the third D-558-1 for extensive investigations of transonic aerodynamic phenomena and aircraft behavior, retiring it in 1953. The D-558-1 #1 did exceed Mach 1 in a 35-degree dive on September 29, 1948, while piloted by Eugene F. May. One Skystreak, the D-558-1 #2, crashed on May 3, 1948, during a takeoff accident following engine failure, killing Howard C. Lilly, a NACA research pilot.

The first D-558-2 Skyrocket completed its initial flight on February 4, 1948, piloted by John F. Martin. The second Skyrocket was used by the NACA to investigate the behavior characteristics of sweptwings. During this program, before its conversion to all-rocket propulsion, the D-558-2 #2 revealed the tendency of swept-wing aircraft to pitch up under certain aerodynamic conditions. After modification in 1950 to all-rocket configuration, the D-558-2 #2 attained Mach 2 and set an unofficial altitude record of 25,370 m (83,235 ft.). All three Skyrockets were retired from flight operations in 1956.

The D-558-1 and D-558-2 were constructed of mixed aluminum and magnesium. Both featured jettisonable nose sections to serve as emergency escape capsules. The first two D-558-1 Skystreaks were bright glossy red overall, but the D-558-1 #3 and later D-558-2 Skyrockets were glossy white, which proved more desirable for optical tracking purposes.

The aircraft was transferred from the US Navy in 1960.

THE DOUGLAS DC-3 was the first airplane to make flying passengers profitable. It was so successful that in 1938, 95 percent of all US commercial airline traffic was on DC-3s, and they are still in use today throughout the world.

In 1933 the Douglas Aircraft Company designed a new passenger plane, as ordered by Transcontinental and Western Air, to compete with the Boeing 247. The first model, the DC-1, was soon succeeded by the DC-2. American Airlines needed a new airplane able to compete with the DC-2 and the Boeing 247, but one with sleeping accommodations.

The new Douglas Sleeper Transport (DST) was a luxury plane with seven upper and seven lower berths and a private forward cabin. The day plane version, known as the DC-3, had twenty-one seats instead of the berths. The design included cantilever wings, all-metal construction, two cowled Wright SGR-1820 1,000-hp radial engines, retractable landing gear, and trailing-edge flaps. The controls included an automatic pilot and two sets of instruments.

American Airlines initiated DST nonstop New York–Chicago service on June 25, 1936, and in September started service with the DC-3. Outclassing the Boeing 247, the DC-3 was faster and carried twice as many passengers.

KLM was the first European airline to own and operate DC-3s in 1936, and they were followed by airlines in Sweden, Switzerland, France, Belgium, and elsewhere, eventually being flown by thirty foreign airlines by 1938. In July 1936, President Franklin D. Roosevelt presented Donald W. Douglas, head of Douglas Aircraft, with the Collier Trophy, recognizing the DC-3 as an outstanding commercial plane.

1. NASM's Eastern Air Lines Douglas DC-3 hanging on display in the Hall of Air Transportation. 2. A line of passengers prepare to embark on an American Airlines Douglas DC-3 in the late 1930s. 3. Passengers prepare for a meal onboard a United Air Lines DC-3. 4. A line of seven C-47 airplanes unloading cargo at Tempelhoff Airport, Berlin, Germany, during the Berlin Airlift.

DIMENSIONS:

WINGSPAN	28.95 M (95 FT.)
LENGTH	19.66 M (64 FT., 6 IN.)
HEIGHT	5 M (16 FT., 11 IN.)
WEIGHT	11,430 KG (25,200 LB.)

The DC-3 was called on to aid the military fleets in World War II. Many commercial carriers in Europe put their DC-3s to use as military transports. The United States ordered new versions of the DC-3 modified for troop transport and cargo carrying. Designated as C-47s and C-53s, the military versions were put into operation in the European and Pacific Theaters. C-47s initiated the Berlin Airlift in 1948.

In England the DC-3 is called Dakota, or Dak. During World War II, American pilots called it the Skytrain, Skytrooper, Doug, or Gooney Bird.

The DC-3 is still in use today throughout the world. Since 1935, 455 commercial transports and 10,174 military versions have been built. In addition, about 3,000 have been constructed under license in Russia (designated Li-2) and almost 500 in Japan.

THE DOUGLAS SBDs accounted for most of the damage from the air sustained by the Japanese in the Battle of Midway, which is regarded by many as the turning point of the war in the Pacific.

Accepted by the US Navy in February 1939, six months before the start of World War II, the Douglas SBD (Scout Bomber Douglas) was a compact, rugged dive-bomber that could take a lot of punishment.

The original navy contract called for 57 SBD-1s; this number increased to 500 after the attack on Pearl Harbor. And, based on lessons learned in the European phase of the war, improvements to the original design included the change to a more powerful engine, a Wright Cyclone R-1820-52, which delivered 1,000 hp to maintain the performance in spite of added weight of the modifications.

The armament had changed: two .50-caliber machine guns had replaced the .30-caliber guns fitted to the earlier planes in the series. A second .30-caliber gun was added to the flexible mount in the rear cockpit.

SBD-2s and SBD-3s played a major role in the crucial battles of the Coral Sea and Midway. In the Battle of Midway on June 4, 1942, SBDs from the USS *Yorktown* attacked the Japanese carrier *Kaga*. This was followed almost immediately by the SBDs from the USS *Enterprise* on the *Akagi* and the *Soryu*. The Japanese lost four large aircraft carriers: the *Kaga, Akagi, Hiryu,* and *Soryu*. The SBD losses were moderate, amounting to thirty-five navy and eight marine planes.

DIMENSIONS: —

WINGSPAN	12.66 M (41 FT., 6 IN.)
LENGTH	10.06 M (33 FT.)
HEIGHT	4.15 M (13 FT., 7 IN.)
WEIGHT	2,967 KG (6,554 LB.)

1. The museum's Douglas SBD-6 Dauntless hanging on display in the Sea-Air Operations gallery. 2. Three US Marine Corps SBD-1 Dauntless members of Scout Bomber Squadron 132 from the Quantico, Virginia, flying in left echelon formation, circa 1941. 3. A US Marine Corps Douglas SBD-2 Dauntless in low-level flight, coming in for a practice landing at an unidentified airfield, June 17, 1942. 4. A Douglas SBD-5A Dauntless in flight. Note the gunner's twin-mounted .30-caliber machine guns.

The SBDs served longer than intended, for their successor, the Curtiss SB2C, was long delayed by modifications required to make them acceptable for carrier service. Unlike most of its contemporaries, the SBD did not have folding wings to improve its shipboard stowage ability.

Ordnance for the SBD could consist of a variety of loads, including a 1,600-pound bomb mounted on the center rack and two 100-pound bombs on wing-mounted racks, all externally mounted. The bomb carried on the center rack was dropped clear of the propeller by a displacing gear, which ensured adequate clearance.

In scouting configuration, drop-tanks could be attached to the wing mounts for greater endurance. Unlike most of its contemporaries, the SBD did not have folding wings to improve its shipboard stowage ability. Instead, it had the same basic wing construction as its forerunners, the Northrop XBT-2, the Northrop Gamma, and the Douglas DC-3.

The National Air and Space Museum's SBD-6 was transferred to the Smithsonian in 1961. Its colors and markings are of aircraft 109 of VS-51, which served on the USS *San Jacinto*.

1

WITH THE SUCCESSFUL crossings of the Atlantic in 1919 by the US Navy's NC-4 and the British flyers, John Alcock and Arthur Whitter Brown, in a Vickers Vimy, the ambition to circumnavigate the globe by airplane was a natural next challenge.

In July 1923 the US War Department disclosed they were sending two officers on an information-gathering trip to stake out a route for a global flight to be attempted by the US Army Air Service. Maj. Gen. Mason M. Patrick, chief of the Air Service, was put in charge of the flight.

For the flight, the Air Service commissioned the Douglas Aircraft Company to design a sturdy, 15-m (49-ft.) span, two-place biplane powered by a single twelve-cylinder, water-cooled 400-hp Liberty engine. Five were built, and they were called Douglas World Cruisers

One of the planes was a prototype used for testing, and the other four—the *Seattle, Chicago, Boston,* and *New Orleans*—were going to be flown around the world as a team. Elaborate preparations were made for fueling and repair sites at strategic locations along the route, arranging overflight and landing clearances, and securing the cooperation of the US Navy and the Royal Air Force. Because the United States did not recognize the Soviet Union at this time, flying over Siberia was prohibited, necessitating a southeast Asian route that added 11,000 km (6,875 mi.).

Shortly after setting off from Seattle on April 6, 1924, the *Seattle* was forced down due to engine trouble. The engine was repaired, but the *Seattle* crashed into a mountainside after getting lost in fog. The pilot and mechanic escaped with minor injuries, but the airplane was destroyed.

DIMENSIONS:

WINGSPAN	15.24 M (50 FT.)
LENGTH	20.66 M (35 FT., 6 IN.)
HEIGHT	4.15 M (13 FT., 7.5 IN.)
WEIGHT	1,987 KG (4,380 LB.)

1. The museum's Douglas World Cruiser DWC-2 *Chicago* on display in the Pioneers of Flight gallery. 2. The Douglas World Cruiser DWC-2 *Chicago* in flight with floats attached to the aircraft. 3. A view of Douglas World Cruiser DWC-4 *New Orleans,* fitted with floats, being launched into the water from a boat ramp, at Reykjavik, Iceland, 1924. 4. Two of the Douglas World Cruisers at anchor in Shanghai, China, surrounded by various junks and other small local vessels.

The three remaining World Cruisers continued on via the northern Pacific chain of islands en route to Japan. They then traversed China, French Indo-China (Vietnam), Siam (Thailand), Burma (Myanmar), India, Persia (Iran), Asia Minor (Turkey), the Balkans, and France. From Strasbourg, they were escorted to Paris by the French Air Force, where they received a tumultuous welcome from cheering crowds on Bastille Day (July 14). The next day they left Paris and landed in London.

Over the North Atlantic between the Orkney and Faroe Islands, the *Boston* suddenly lost oil pressure and had to alight in the ocean. Although the landing was successful, the *Boston* was damaged beyond repair during an attempt to hoist it on board a navy ship. At Pictou, Nova Scotia, the prototype World Cruiser joined the remaining two and became the *Boston II.* From there the planes flew on for a triumphal journey across the United States, arriving in Seattle on September 28. The 44,085-km

(27,553-mi.) flight was completed in 175 days with a flying time of 371 hours and 11 minutes at an average speed of 112 kph (70 mph).

The 1924 round-the-world flight remains one of the truly great achievements in aviation. It was an incredibly arduous trek with an endless series of forced landings, repairs, bad weather, and other mishaps. It was a monumental logistical accomplishment and an important step toward worldwide air transport in the decades to come.

Two weeks before the Douglas World Cruisers completed their flight around the world, a young museum aide named Paul E. Garber recommended that the Smithsonian acquire one of the aircraft for its collection. Eleven months later the secretary of war approved the transfer of the *Chicago* to the Smithsonian. The *Chicago* was restored between 1971 and 1974 and moved into the new National Air and Space Museum in 1976.

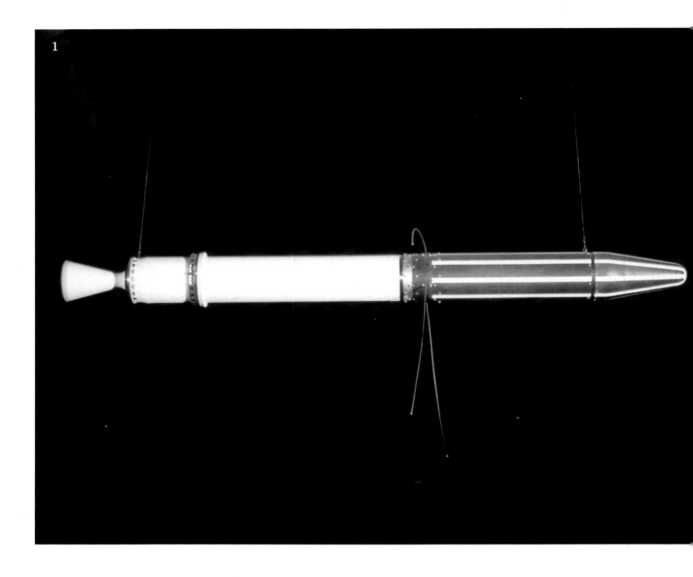

1

DATA TRANSMITTED FROM the Explorer satellites led to the discovery of a belt of intense radiation that surrounds the Earth.

The *Explorer 1* satellite was placed into Earth orbit on January 31, 1958, becoming the first artificial satellite to be launched successfully by the United States. Known unofficially as Satellite 1958 alpha, *Explorer 1* transmitted data on micrometeorites and cosmic radiation for 105 days.

The payload section consists of a bullet-shaped 203-cm (80-in.) long, 15-cm (6-in.) diameter cylinder with an aerodynamic nose cone. This is attached to the fourth stage of the Jupiter C launch rocket at the antenna ring. Four whip antennas project from that antenna ring at the payload-rocket interface. Instruments aboard the satellite included a halogen-quenched Geiger counter to detect cosmic rays and three micrometeorite detectors (a microphone, a wire grid detector, and a metallic film detector). Data were radioed to ground stations by two mercury-battery-powered transmitters from a 10 mW FM and a 60 mW AM radio.

Assembled by the Jet Propulsion Laboratory at Caltech in just three months, *Explorer 1* was launched on January 31, 1958, from Cape Canaveral. The satellite went into an orbit with a perigee of 356 km (221 mi.) and an apogee of 2,548 km (1,538 mi.) and an

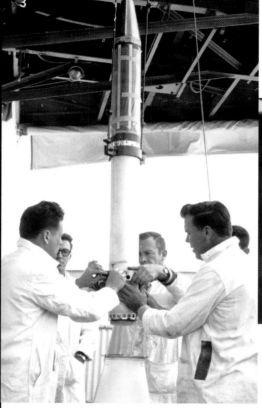

DIMENSIONS:

LENGTH   203 CM (80.75 IN.)
DIAMETER   15 CM (6.5 IN.)
WEIGHT   14KG (31 LB.)

1. A side view of the museum's back-up *Explorer 1* satellite hanging on display in the Milestones of Flight gallery. **2.** A dramatic night launch of the *Explorer 1* satellite on board the Jupiter C rocket at Cape Canaveral, Florida, on January 31, 1958. **3.** Technicians fit the *Explorer 1* satellite to the top of a Jupiter C rocket in January 1958. **4.** A model of the *Explorer 1* being jubilantly held up by William Pickering, James Van Allen, and Wernher von Braun, at the National Academy of Science, Washington, DC, January 31, 1958.

inclination from the equator of 33.24 degrees. There was no onboard data-storage capability, and the data could only be received when the satellite passed within range of one of the ground stations.

Data from the Geiger counter surprisingly showed a total lack of particle counts above altitudes of 1,000 km (621 mi.) leading at first to suspicions of instrument failure. When *Explorer 3* made the same observation, it was

determined that there was a belt of intense radiation around the Earth. This belt was named after its discoverer, James Van Allen. *Explorer 1* continued to transmit data until May 23, 1958, when its batteries finally died.

The *Explorer 1* on display in the National Air and Space Museum's Milestones of Flight gallery is a spare satellite built at JPL and transferred to the Smithsonian in 1961.

AN IMPORTANT STEPPING-STONE to space exploration, the *Explorer II* balloon flight conclusively demonstrated that humans could survive and work in a pressurized environment high above the Earth.

The Explorer project was conceived by Capt. Albert Stevens, chief of the Army Air Corps photography laboratory at Wright Field, Ohio. With funding from the National Geographic Society, Maj. William E. Kepner, Capt. Orvil A. Anderson, a noted Army Air Corps aeronaut, and Stevens attempted a world-altitude–record flight in 1934 with *Explorer I*. Known as the Stratobowl, the launch site was a natural depression approximately 91 to 152 m (300 to 500 ft.) below the surrounding tree-lined limestone cliffs in the Black Hills of South Dakota. The floor of the hollow covered .14 sq km (35 acres) and was ideally suited for the launching of balloons, because it provided natural protection from winds.

When the *Explorer I* balloon ripped shortly after launch, its hydrogen mixed with air

DIMENSIONS: ══════

DIAMETER   2.8 M (9 FT.)
WEIGHT   290 KG (640 LB.) EMPTY

1. The *Explorer II* gondola on exhibit in the museum's Pioneers of Flight gallery. **2.** The fully inflated balloon *Explorer II* in place at the center of the Stratobowl, near Rapid City, South Dakota, five minutes before its ascent on its stratospheric research flight, November 11, 1935. **3.** Capt. Albert W. Stevens and Capt. Orvil A. Anderson, of the National Geographic–Army stratosphere expedition in the instrument-filled gondola.

and exploded. After a harrowing few moments the crew escaped through a hatch and parachuted to safety.

For *Explorer II,* the hatch was widened for easier escape, and the balloon was filled with 104,772 cu. m (3.7 million cu. ft.) of helium. This was the first such balloon to use this inert gas. The balloon was enlarged, the crew was cut from three to two, and the weight of its scientific payload was halved. The balloon envelope was constructed of rubberized fabric.

Like that of *Explorer 1,* the *Explorer II* gondola was constructed of welded magnesium/aluminum alloy sections. The gondola was sealed and pressurized to protect the aeronauts from the intense cold and insufficient oxygen needed to sustain human life in the stratosphere. The 2.8-m (9-ft.) sphere weighed 290 kg (640 lb.) and carried a payload of 700 kg (1,500 lb.). The balloon reached its maximum altitude of 22 km (14 mi.) and remained airborne for 8 hours and 13 minutes covering 363 km (225 mi.) over the ground. Anderson and Stevens

broadcast a running commentary of their flight through a shortwave radio.

Captain Stevens was able to photograph the division between the troposphere and the stratosphere and the actual curvature of the Earth. His cameras also captured South Dakota and surrounding states, and demonstrated the potential of high-altitude, long-range reconnaissance from manned balloons.

While they carried fewer instruments than on the previous flight, Anderson and Stevens were able to collect much useful data about the stratosphere from the sixty-four scientific devices they did bring. The instruments aboard *Explorer II* collected data about cosmic rays, the ozone layer, aeronomy, meteorology, biology, radio propagation in the high atmosphere, the effect of high altitude on insects, and microorganisms in the stratosphere.

Explorer II was such a success that Stevens and Anderson were awarded the prestigious Harman and MacKay Trophies. It was transferred from the Army in 1937.

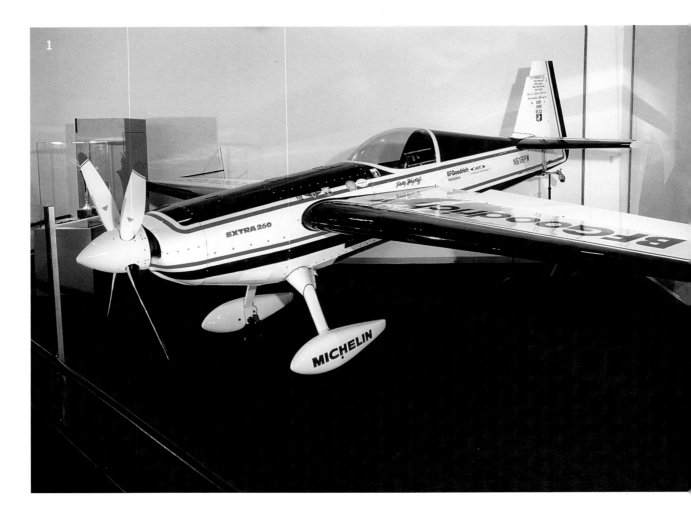

FLYING THE EXTRA 260, Patty Wagstaff became the first woman to win the US National Aerobatic Championship in September 1991.

The Extra 260 is a one-of-a-kind aircraft created by Walter Extra, a former German aerobatic competitor and one of the world's premier aerobatic aircraft designers and builders. Recognized for its beauty, high performance, and maneuverability, it can roll at the rate of 360 degrees per second and climb vertically at 1,200 m (4,000 ft.) per minute.

The Extra 260, built by Extra's firm, Extra Flugseugbau, is a successful blend of traditional and high-technology construction. The steel-tube fuselage and wings, made of Polish pine box spar and solid ribs, and covered in birch plywood, are then covered with Ceconite fabric. These standard aircraft construction materials contrast with the modern composite material used for the horizontal and vertical tail surfaces and landing gear.

Featuring an almost full-length, carbon fiber aileron, the wing is designed to give the aircraft lateral instability and permits slipping and skidding, unlike most other aircraft. This instability results in greater maneuverability and faster response to the pilot's touch. Each wingtip has a sighting device to aid in precision maneuvering.

The aircraft's Textron Lycoming AEIO-540-D4A5 engine is a modified version of the standard 540 high-performance, six-cylinder engine used in many general aviation aircraft. The modifications allow the Extra 260 to make complex maneuvers, such as multiple vertical snap rolls and knife-edge flight. Wagstaff continually ran the engine at or above the "red line" (maximum rpm) during competitions and air shows.

WINGSPAN	7.6 M (24 FT., 9 IN.)
LENGTH	6 M (20 FT.)
HEIGHT	1.8 M (6 FT.)
WEIGHT	2,530 KG (1,150 LB.)

1. Patty Wagstaff's Extra 260 on display in the museum's Pioneers of Flight gallery.
2. The museum's Extra 260 in a straight vertical climb. **3.** Patty Wagstaff poses in front of her Extra 260.

Patty Wagstaff received her private pilot's license in Alaska in 1980 and began aerobatic instruction in 1983. She moved quickly into competition flying and advanced to the Unlimited category of competition in only two years. In 1986 she qualified as a member of the US aerobatic team and competed in six biennial World Aerobatic Competitions. Winning three National Championships, medals in world competition, and numerous trophies, she was a dominating force in aerobatic contests until her retirement from competition in 1996.

Wagstaff's family donated the Extra 260 to the National Air and Space Museum in 1993.

1

IN 1937 THE German Air Ministry issued a contract to Focke-Wulf Airplane Company for a single-engine plane to supplement the standard Luftwaffe fighter, the Messerschmitt Bf 109.

The prototype, designated Fw 190 V1, took to the air on its maiden flight on June 1, 1939, and was the only German fighter of World War II that flew with a radial engine. It had excellent performance and handling but high engine and cockpit temperatures, which stemmed from a too tight cowling design that choked airflow around the engine.

With some adjustments, the Fw 190A-1 entered service with JG 26 (Jagdgeschwader, or Fighter Wing) in France during August 1941. In September pilots flying the new Focke-Wulf tangled with Spitfires, and the Allied fighter proved inferior except in turning radius. Until the improved Spitfire Mk. IX was introduced in late 1942, the Allies had no fighter to equal the Focke-Wulf.

A BMW 801D-2 engine, capable of producing 2,100 hp for brief periods by using a methanol-water injection system called MW-50, powered the Fw 190A-4. The engine was moved forward 15 cm (6 in.) on the the A-5, which finally solved the cooling problem.

Fw 190A-5–equipped Luftwaffe fighter units inflicted heavy losses on unescorted US heavy bombers during 1943. The A-7 and the A-8 (the fighter version produced in the greatest numbers) incorporated heavier armament that proved devastating against Allied

DIMENSIONS:

WINGSPAN	10.5 M (34 FT., 5.5 IN.)
LENGTH	8.84 M (29 FT.)
HEIGHT	3.98 M (13 FT.)
WEIGHT	3,170 KG (7,000 LB.)

1. A view of NASM's restored Focke-Wulf Fw 190F-8 taken October 11, 1983, at the Paul E. Garber Facility. 2. A captured Focke-Wulf Fw 190G-3 in flight over Ohio. 3. A Focke-Wulf Fw 190F on the ground at Villafranca, Italy, in the custody of the 57th Fighter Group, May 1945. This aircraft was landed at the Villafranca airfield by a German pilot who was unaware that the airfield had been captured by Americans. 4. Three Focke-Wulf Fw 190A fighters on the line with engines running.

bombers, but it also added weight. The Fw 190 became more vulnerable to US escort fighters such as the Republic P-47 Thunderbolt and North American P-51 Mustang.

The Fw 190D carried a powerful, liquid-cooled Junkers Jumo 213 engine. It proved to be an excellent fighter but arrived too late in the war to compensate for fuel shortages and inexperienced pilots.

The 190's reliable, air-cooled engine and wide-track landing gear were ideally suited to the harsh conditions of the Eastern Front. Operations there led to several new variants, such as the Fw 190F fighter-bomber, which carried 360 kg (794 lb.) of armor, including sections of steel plate behind the pilot's head, on the lower engine cowling, and on the wheel-well doors. The F-8 model became the most important variant of the entire F series. Using kits supplied by the factory, frontline units adapted

these airplanes to carry various combinations of heavy cannons, bombs, rockets, and even torpedoes.

The Fw 190 in the National Air and Space Museum left the production line in late 1943 as an Fw 190A-7 fighter. After suffering damage, it was repaired and remanufactured into an Fw 190F-8 fighter-bomber. The conversion involved fitting a new wing and bomb racks to the original fuselage and adding armor plate around and beneath the cockpit. The reissued aircraft flew on the Eastern Front in late 1944, probably based in Hungary.

Captured by the Allies before the war ended, the US Air Force had it transferred to the Smithsonian in the 1950s. The final paint and markings applied are historically accurate for the SG 2 (Schlagtgeschwader, or Ground-Attack Squadron 2) that fought in Hungary.

DURING THE LATTER half of 1917, the Allies had regained air superiority over the Western Front with the S.E.5 and the Spad fighters. To counter this, the German government invited aircraft manufacturers to submit prototype single-seat fighter designs for evaluation at a competition at the Adlershof airfield in Berlin in January 1918.

One of the winners was the Fokker V.11, the D.VII's prototype, offered by the Dutch-born aircraft manufacturer, Anthony Fokker.

The V.11 was largely the creation of Fokker's chief designer, Reinhold Platz. Platz was the true creative force behind the famous Fokker fighters of the second half of the war. Anthony Fokker's talents were greater as a test pilot than as a designer. Fokker's ego and dominating personality frequently led him to understate Platz's role as the genuine innovator of the designs that bore the Fokker name. Nevertheless, his intuitive piloting and political skills combined with Platz's innovative preliminary designs, made for a formidable team. This was especially true in the case of the Fokker D.VII.

The Fokker V.11 was completed just before the Adlershof competition. German ace, Manfred von Richthofen, "the Red Baron," tested the V.11 at Fokker's request. Richthofen thought the airplane had generally good performance but was tricky to handle and directionally unstable, especially in a dive. To remedy these problems, Fokker lengthened the fuselage 40 cm (16 in.), added a fixed vertical fin and a new rudder shape, and altered the aileron balances, among other small changes. With these modifications, the V.11 was safe and pleasant to fly, and had lost little of the maneuverability that had initially impressed von Richthofen.

2   3

4

1. The museum's Fokker D.VII "U.10" on display in the World War I gallery, circa 1980. **2.** A period image of the museum's Fokker D.VII "U.10" following its capture by US troops on November 9, 1918. American soldiers have added the "kicking mule" insignia to the airplane. **3.** A Fokker D.VII pilot is posed by the cockpit of his aircraft. **4.** An informal portrait of Lt. Ulrich Neckel, commander of Jasta 6, 30-victory ace. Neckel is in the cockpit of his Fokker D.VII. Notice the toy teddy bear attached to the rearview mirror on the upper wing.

DIMENSIONS: _____

WINGSPAN	8.93 M (29 FT., 3.5 IN.) UPPER;
	6.86 M (22 FT., 10 IN.) LOWER
LENGTH	7.01 M (23 FT.)
HEIGHT	2.82 M (9 FT., 3 IN.)
WEIGHT	700 KG (1,540 LB.)

After winning the Adlershof competition, Fokker received a production order for 400 Fokker D.VIIs. Concerned that the Fokker factory was unable to meet the demand for the new fighter, Albatros, a rival of Fokker, was directed to produce D.VIIs. Because the Fokker factory produced no construction drawings and working only from jigs and assembly sketches, Albatros had to make its own drawings based on a completed airframe. The result was Fokker- and Albatros-built D.VIIs that looked alike, but differed in detail. Albatros produced more D.VIIs than Fokker (about 1,000) and they were generally considered to be of higher quality. Also, not all of the components were interchangeable between the D.VIIs of the two manufacturers.

Fokker D.VIIs began to reach frontline units in April 1918. Initially, the D.VII was powered by a 160-hp Mercedes D.III engine. By the summer, however, the Mercedes-powered D.VII was having difficulty keeping pace with the latest Allied fighters. The airplane was then fitted with the new 185-hp BMW IIIa, which dramatically improved performance. When the Fokker D.VII appeared on the Western Front, Allied pilots at first underestimated the new fighter because it lacked the sleek, graceful lines of the Albatros fighters. But they revised their view after realizing that the Fokker D.VII had the ability to seemingly "hang on its propeller" and fire into the unprotected underside of Allied two-seater reconnaissance aircraft.

The Fokker D.VII's thick-wing section endowed the airplane with good stall characteristics. Positioning themselves below and behind a two-seater where the enemy observer couldn't bring his guns, a D.VII pilot could safely put his airplane into a nose-high attitude in near stall. They were so effective that the Armistice agreement specifically demanded that all Fokker D.VII aircraft should immediately be surrendered.

The Fokker D.VII in the National Air and Space Museum is Albatros-built. It was given to the Smithsonian by the War Department in 1920 and fully restored in 1961.

THE FOKKER T-2 was the first airplane to make a nonstop flight across the United States.

The product of famed Dutch manufacturer Anthony Fokker and his chief designer, Rheinhold Platz, the Air Service Transport 2, or T-2, was built in the Netherlands in 1922. It was the fourth in a series of commercial transport designs by the Fokker Company.

The largest Fokker aircraft up to that time, the T-2 featured a fully cantilevered wooden monoplane wing spanning nearly 25 m (82 ft.) and a fuselage just short of 15 m (49 ft.) long. It was powered by an American-built 420-hp Liberty V-12 engine. In its standard configuration, the Fokker had a single pilot's position located in a forward open cockpit to the left side of the engine. The enclosed cabin carried eight to ten passengers and their baggage.

Early trials by the US Air Service indicated that the T-2 was capable of carrying heavy loads and could be adapted to make the long-distance flight from coast to coast. The center section of the wing had to be reinforced to handle the added weight resulting from the greatly increased fuel supply. The standard 492-l (130-gal.) fuel tank, located in the leading edge of the wing, was supplemented by a 1,552-l (410-gal.) tank in the wing center section and a 700-l (185-gal.) tank mounted in the fuselage cabin area. Also installed in the cabin was a second set of controls for when the two-man crew exchanged positions.

After two unsuccessful attempts starting from the West Coast, Lts. Oakley G. Kelly and John A. Macready took off from the Roosevelt-Hazelhurst Field on Long Island, New York, at 12:30 p.m. on May 2, 1923. At takeoff, the airplane had a gross weight of 4,932 kg (10,850 lb.)—only 68 kg (150 lb.) less than the T-2's specified limit.

DIMENSIONS:

WINGSPAN	24.26 M (79.57 FT.)
LENGTH	15 M (49 FT., 3 IN.)
HEIGHT	3.71 M (12 FT., 2 IN.)
WEIGHT	4,922 KG (10,850 LB.)

3

4

1. The museum's US Army Air Service Fokker T-2 (s/n A.S.64233) on display in Pioneers of Flight gallery. **2.** A view of the Fokker T-2 in flight. **3.** Lt. John A. Macready (left) and Lt. Oakley G. Kelly (right) pose with gasoline barrels and cans arranged beneath the Fokker T-2. **4.** The crew of the Fokker T-2, Lt. John A. Macready (left) and Lt. Oakley G. Kelly stand in front of the aircraft in which they made the first nonstop flight across the US on May 2–3, 1923.

Kelly was at the controls first, and flew as far as Richmond, Indiana, and then switched places with Macready, who flew until midnight, at which time they were approaching the Arkansas River, the 1,900 km (1,188 mi.) point. They exchanged positions there; again at Santa Rosa, New Mexico, at 6:00 a.m. the following morning; and once more as they crossed the Great Divide at an altitude of 3,110 km (10,200 ft.). Macready landed the T-2 in San Diego on May 3 at 12:26 p.m., local time, completing the nonstop transcontinental journey in an official time of 26 hours, 50 minutes, and 38.6 seconds. The T-2 had flown 3,950 km (2,470 mi.) at an average ground speed of 147 km/h (92 mph).

The Air Service transferred the T-2 to the Smithsonian in 1924. It was fully restored in preparation for display in the new National Air and Space Museum building in 1976.

ONE OF THE MOST important events in the selling of aviation to the general public was the entry of Henry Ford into aircraft manufacturing. The Ford Motor Company was a symbol of safety and reliability, and the Ford Tri-Motor was a rugged, dependable transport airplane that won a permanent place in aviation history.

In April 1925 the Ford Motor Company started an experimental air freight service between Detroit and Chicago, and in August, Ford purchased the Stout Metal Airplane Company. Up to this point, Stout airplanes used a single engine. The introduction of the lightweight Wright air-cooled radial engine, however, set Stout and his design team onto a new course: a three-engine airplane.

The first Ford Tri-Motor was retroactively designated 3-AT (for air transport). It was an unsightly airplane, which could not be landed power-off because of the terrible airflow patterns generated by its unusually positioned engines. A mysterious fire broke out in the factory in January 1926, after the third flight of the 3-AT, destroying that airplane.

A team of engineers began work on the improved 4-AT, which made its maiden flight on June 11, 1926. By the time Ford stopped producing aircraft in 1933, 199 Tri-Motors had been built. Ford had two overwhelming advantages over the domestic market: the Ford name and all-metal construction. More than one hundred airlines flew the Ford in the United States, Canada, Mexico, Central and South America, Europe, Australia, and China.

Increasing airline use and the availability of the new Pratt & Whitney 420-hp Wasp engine led to the 5-AT model in the summer of 1928, the most famous of the Ford Tri-Motor designs. Two other types, the 8-AT and 14-AT, did not get beyond prototypes.

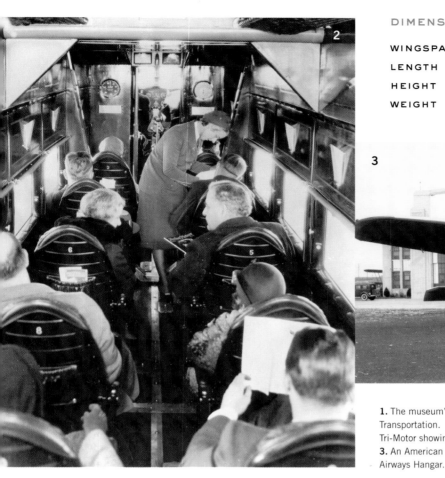

DIMENSIONS: ═══

WINGSPAN	23.71 M (77 FT., 10 IN.)
LENGTH	15.18 M (49 FT., 10 IN.)
HEIGHT	4.16 M (13 FT., 8 IN.)
WEIGHT	3,470 KG (7,650 LB.)

1. The museum's Ford 5-AT Tri-Motor hanging on display in the Hall of Air Transportation. **2.** The passenger cabin of a United Air Lines Ford 5-AT Tri-Motor showing seated passengers; cockpit visible at far end.; circa 1932. **3.** An American Airways Ford 5-AT-B Tri-Motor in front of the American Airways Hangar.

The Ford Tri-Motor is an inherently stable airplane, designed to fly well on two engines and to maintain level flight on one. Its rugged construction and ability to operate from grass and dirt airstrips kept the Tri-Motor in operation for many years.

The Tri-Motor in the National Air and Space Museum is a 5-AT-B donated by American Airlines. Its long and varied history began when it was sold to Southwest Air Fast Express (SAFE) on April 12, 1929, for $55,475 cash. American Airways bought out SAFE the following year, acquiring the Tri-Motor to fly the route between Cleveland and Los Angeles.

In 1936 the airplane was sold to TACA International Airlines and operated in Nicaragua for several years. In 1946 the Tri-Motor was sent to Mexico, where it was used for passenger and cargo hauling until 1954, when it was sold to a crop-dusting company in Montana. It finally ended up beside a small airfield in Oaxaca, Mexico, as someone's living quarters. A wood-burning stove had been installed, and a chimney stuck through the aluminum roof.

Reacquired by American Airlines, the 5-AT was fully restored and flown on public relations tours throughout the country. They donated it to the museum in 1973, where it hangs in the Air Transportation gallery.

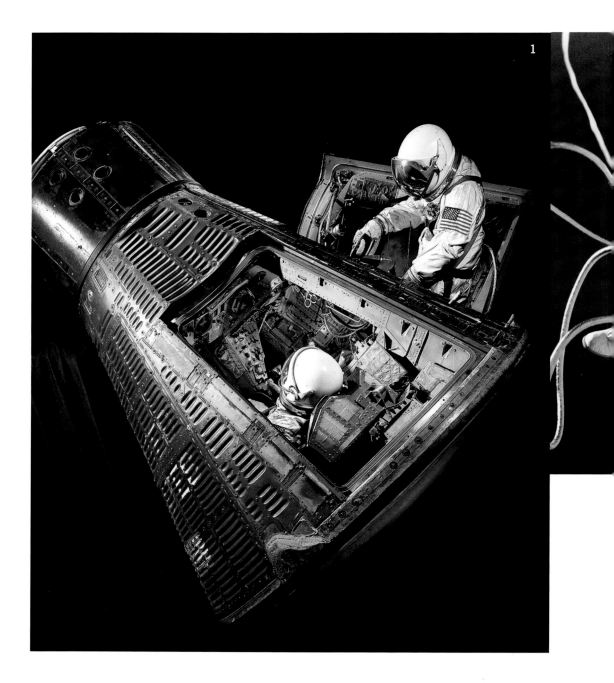

1

ON JUNE 3, 1965, astronaut Edward H. White II became the first American to perform a spacewalk, or extravehicular activity (EVA).

During his 22-minute spacewalk, White remained connected to Gemini 4's life-support and communications systems by the golden "umbilical cord," and he used a handheld jet thruster to maneuver in space. His crewmate, James A. McDivitt, remained inside the spacecraft. The first EVA had been performed three months earlier by Soviet cosmonaut Aleksei A. Leonov, who remained outside his spacecraft for about 10 minutes.

Gemini 4 (GT-4) was launched into Earth's orbit aboard a Titan II rocket on June 3, 1965. The flight lasted four days and White's historic spacewalk was broadcast live. The flight also included a rendezvous maneuver with the second stage of the Titan II rocket. The maneuver was aborted after pilot McDivitt experienced unexpected difficulties catching

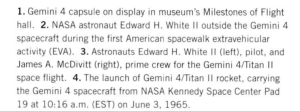

DIMENSIONS:

HEIGHT	3.4 M (11 FT.)
MAXIMUM DIAMETER	2.3 M (7 FT., 6 IN.)
WEIGHT	3,200 KG (7,000 LB.)

1. Gemini 4 capsule on display in museum's Milestones of Flight hall. **2.** NASA astronaut Edward H. White II outside the Gemini 4 spacecraft during the first American spacewalk extravehicular activity (EVA). **3.** Astronauts Edward H. White II (left), pilot, and James A. McDivitt (right), prime crew for the Gemini 4/Titan II space flight. **4.** The launch of Gemini 4/Titan II rocket, carrying the Gemini 4 spacecraft from NASA Kennedy Space Center Pad 19 at 10:16 a.m. (EST) on June 3, 1965.

up to the booster stage. Other experiments during this flight involved photographing the Earth, measuring space radiation, and collecting data on the medical effects of prolonged weightlessness.

Gemini 4 was the second of ten manned Gemini missions, which perfected the techniques of spacecraft rendezvous and docking, demonstrating that astronauts could withstand prolonged weightlessness. The data collected in these missions was used for the Apollo missions to the Moon.

The configuration in the National Air and Space Museum is the only part of Gemini 4 that returned to Earth. Behind the heat shield was an adapter section containing propellants for the maneuvering thrusters, fuel cells for electric power, and retrorockets to enable the return to Earth. The heat shield was jettisoned before reentry. The nose section was discarded during deployment of the main parachute, and the spacecraft landed on the ocean with the hatches facing up. NASA transferred Gemini 4 to the museum in 1967.

ON MARCH 16, 1926, in Auburn, Massachusetts, Robert H. Goddard launched the world's first successful liquid-fueled rocket. The flight reached an altitude of 12 m (41 ft.), lasted 2.5 seconds, and covered a horizontal distance of 56 m (184 ft.).

Funded by the Smithsonian and private foundations, Goddard experimented with liquid-propellant rockets before anyone else. But he worked alone and was reluctant to publicize his results.

Goddard recognized that liquid propellants provided more energy for propulsion than

DIMENSIONS:

LENGTH	6.7 M (21 FT., 11 IN.)
DIAMETER	.47 M (18 IN.)
EMPTY WEIGHT	73 KG (161 LB.)
LOADED WEIGHT	107 KG (236 LB.)

1. A view of the Goddard 1941 liquid-propellant rocket on display in the Milestones of Flight gallery. **2.** Robert H. Goddard stands beside the launch stand of a liquid-fuel rocket prior to its test launch on March 8, 1926. **3.** Dr. Robert H. Goddard and assistants work on the turbopump on its assembly frame in the Goddard shop at Roswell, New Mexico.

gunpowder or other available solid fuels. In his earliest rockets, he placed the engine at the top of the vehicle and the fuel tanks below. However, he soon found that this "nose drive" arrangement was too unstable, so he placed the motor at the bottom. Goddard's liquid-propellant rockets burned liquid oxygen and gasoline.

By February 9, 1940, a full-sized rocket with a pump system was ready to fly. This was designated test P-15, but it failed to lift because the pump clogged with frozen liquid oxygen. On August 9, 1940, the P-23 flew, but extremely poorly. It reached a speed of 16 to 24 km/h (10 to 15 mph) as it cleared the tower then tipped over and reached about 61 to 91 km (200 to 300 ft.) before crashing into the ground with a tremendous explosion.

The most sophisticated of Goddard's rockets was the 1941 P-series, which had turbine-powered pumps and incorporated other features he had worked out over the years, including a high-efficiency combustion chamber, gyroscopic stabilization, and blast vanes for correcting the course during flight. All of his later rockets were equipped with parachutes and a parachute-release system for

recovery so that he could study how things went and also reuse undamaged hardware.

In test P-31 on May 8, 1941, a rocket flew up about 76 km (250 ft.) before it tipped over and exploded. That was Goddard's last successful flight, though there were other attempts and static tests.

In 1942, Goddard and his small team transferred to Annapolis, Maryland, to help the navy develop controllable liquid-propellant JATO (jet-assisted takeoff) rockets for heavily loaded seaplanes, where he worked until his death in 1945.

Despite his failures and not sharing his work with others, Goddard's accomplishments are still considered remarkable. He had developed many of the same rocketry concepts as the Germans, who had independently developed the V-2 missile used during World War II, though he had far fewer financial resources and a much smaller technical team.

The rocket in the National Air and Space Museum is probably the one he used for test P-36 on October 10, 1941, which jammed in the launch tower and failed to lift. It was donated to the Smithsonian in 1950.

THE F4F-4 WILDCAT fighter held the line in the Pacific for the first two years of World War II, although often outnumbered by the superior Japanese Zero.

Grumman not only sold F4Fs to the US Navy and Marine Corps but also to the French, who were in dire need of aircraft of all kinds. When France capitulated to Germany in 1940, its production contracts were taken over by the British Purchasing Commission. The F4F, called the Martlet by the British, served with the Fleet Air Arm. The Martlet gained the distinction of becoming the first US aircraft in British service to shoot down a German aircraft—a Junkers JU.88—in World War II.

In the Pacific, the Wildcat first flew in defense of Wake Island in December 1941. Marine Fighter Squadron VMF-21 1 lost eight of its twelve F4F-3 Wildcats on the opening day. The remaining four were flown continuously, fighting heroically for two weeks, breaking up many air attacks and sinking a cruiser and a submarine with 45.4-kg (100-lb.) bombs before the last two Wildcats were destroyed the day the Japanese landed on Wake Island.

Despite similar losses throughout the Pacific, this tough fighter had a kill average for the war of nearly 7 to 1. By 1942 all American fighter squadrons were equipped with the folding-wing F4F-4. The Wildcat's basic opponent was the Japanese Zero, a fighter that could outmaneuver and outperform it. The Wildcat's heavy armament and solid construction, however, gave it an advantage in the hands of skilled pilots.

2 **3** **4**

DIMENSIONS: ————

WINGSPAN	11.58 M (38 FT.)
LENGTH	8.77 M (28 FT., 9 IN.)
HEIGHT	4.52 M (9 FT., 2.5 IN.)
WEIGHT	2,612 KG (5,758 LB.)

1. The museum's US Navy Grumman FM-1 (F4F-4) Wildcat on display in the Sea-Air Operations gallery. **2.** The flight deck of a US Navy escort carrier unfold the wings of an F4F Wildcat in preparation for takeoff. **3.** A formation of four Grumman F4F Wildcats in flight. **4.** A Grumman F4F-3A Wildcat of Squadron VF-6 receiving the "go" signal to launch from unidentified US Navy aircraft carrier during World War II.

A new Grumman fighter, the F6F Hellcat, was built as the replacement, but the navy still needed the Wildcat to equip the small escort carriers, for which it was well suited in size and weight. To make room for production of the Hellcat at the Grumman plant, Wildcat manufacture was transferred to the Eastern Aircraft Division of General Motors. These fighters emerged with the designations FM-1 and FM-2.

The Wildcat in the National Air and Space Museum is the 400th FM-1 built. The paint duplicates the US Navy blue-gray camouflage used early in the war, and the markings are patterned after an FM-1, number E-10, that operated from the escort carrier USS *Breton* in the Pacific in mid-1943. The museum received the Wildcat from the navy in 1944.

DURING ITS SHORT but successful career, the popular F6F Hellcat was flown by 305 aces, more than any other American fighter during World War II.

In the first two years of US involvement in the war, the US Navy did not have an aircraft as maneuverable as the Japanese Zero fighter, and the Americans depended on the Grumman F4F Wildcat (see page 80) until a replacement aircraft could be produced.

Wildcat design provided the foundation, but for the new Hellcat, Grumman engineers abandoned a round fuselage cross section for a teardrop-shaped cross section. This streamlined fuselage surrounded the large Pratt & Whitney R-2800-10W Double Wasp engine.

The Hellcat's backward-rotating folding wing had the largest wing area (31 sq. m [334 sq. ft.]) of any US fighter aircraft produced during the war. This large square-panel wing was mounted at the minimum angle of incidence to obtain the least drag in level flight. However, a relatively large angle of attack was required so the Double Wasp engine was mounted with a negative thrust line. This combination resulted in a tail-down flight attitude.

DIMENSIONS:

WINGSPAN	13.04 M (42 FT., 10 IN.)
LENGTH	10.24 M (33 FT., 7 IN.)
HEIGHT	3.99 M (13 FT., 1 IN.)
WEIGHT	4,152 KG (9,153 LB.)

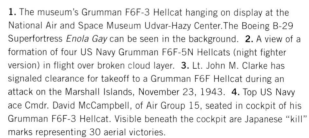

1. The museum's Grumman F6F-3 Hellcat hanging on display at the National Air and Space Museum Udvar-Hazy Center. The Boeing B-29 Superfortress *Enola Gay* can be seen in the background. **2.** A view of a formation of four US Navy Grumman F6F-5N Hellcats (night fighter version) in flight over broken cloud layer. **3.** Lt. John M. Clarke has signaled clearance for takeoff to a Grumman F6F Hellcat during an attack on the Marshall Islands, November 23, 1943. **4.** Top US Navy ace Cmdr. David McCampbell, of Air Group 15, seated in cockpit of his Grumman F6F-3 Hellcat. Visible beneath the cockpit are Japanese "kill" marks representing 30 aerial victories.

The design proved an immediate success. Hellcats first entered combat on August 31, 1943, in a series of raids on Marcus Island. American pilots now had an aircraft that was faster than and almost as maneuverable as the Japanese. The Hellcat's most successful day in combat came on June 19, 1944, during operations in the Mariana Islands. During this air battle that became known as the Great Marianas Turkey Shoot, the Japanese lost over 270 aircraft compared with 26 lost by the United States.

The Hellcat also saw combat in Europe. The Royal Navy's Fleet Air Arm escorted Fairey Barracudas during raids on the German battleship *Tirpitz* hidden in a Norwegian fjord. They were also flown by American pilots over the invasion beaches in southern France during Operation Anvil. At a ratio of 19 to 1, the war pilots flying Hellcats destroyed over 5,100 enemy aircraft, compared with 270 Hellcats lost in combat. By war's end, 12,275 Hellcats had been built.

The F6F-3 Hellcat in the National Air and Space Museum was donated by the US Navy in 1960. The plane is painted with the US Navy's tricolor camouflage and markings of an F6F that served in VF-5 off the USS *Yorktown*.

ONE OF THE MOST exciting aerobatic aircraft of the 1930s and 1940s was Al Williams's *Gulfhawk II*. Grumman Aircraft built the G-22 *Gulfhawk II* for Maj. Alfred "Al" Williams, former naval aviator and marine, who at the time was head of the aviation department at the Gulf Oil Company.

This sturdy civilian biplane, the sole G-22 made, was a cross between the successful F2F and F3F fighters Grumman was building for the US Navy and Marine Corps in the mid-1930s. The G-22 was powered by a Wright Cyclone R-1820-GI 1,000-hp engine and a three-blade Hamilton Standard propeller. Taken from the F2F-1 biplane fighter, the fabric-covered wings of *Gulfhawk II* were of unequal span, constructed of aluminum spars and ribs. The fuselage was from the later XF3F-2 and was of monocoque construction, covered with a 0.08-cm (0.032-in.) aluminum alloy, and accommodated only the pilot.

Williams's G-22 featured the larger vertical tail surfaces of the production F3F-2 but without the ventral fin extension that was fitted to the standard production F3F-2. Already stressed to an exceptionally high 9g load factor, the *Gulfhawk II* was further modified to withstand the even higher load factors encountered during aerobatics. The aircraft was equipped for inverted flying for periods of up to half an hour.

The *Gulfhawk II*'s orange and blue paint represented the Gulf Oil Company, the sponsors of Williams's exhibition flight.

The airplane thrilled many an air show spectator throughout the United States and Europe from 1936 to 1948. It was a feature attraction at such meets as the Cleveland Air Races, the Miami All-America Air Show, and the New York World's Fair, demonstrating precision aerobatics and the new technique of dive-bombing.

WINGSPAN	8.72 M (28 FT., 7 IN.)
LENGTH	7.01 M (23 FT.)
HEIGHT	3.05 M (10 FT.)
WEIGHT	1,625 KG (3,583 LB.)

1. NASM's Grumman G-22 *Gulfhawk II* hanging on display at the Udvar-Hazy Center. **2.** Maj. Alford "Al" Williams poses standing beside Grumman G-22 *Gulfhawk II* during the 12th All-American Air Maneuvers, Miami, Florida, 1939. **3.** The Grumman G-22 *Gulfhawk II* on display during an unidentified air show. **4.** Al Williams at the controls of the Grumman G-22 *Gulfhawk II* in flight.

In 1938, Williams crated and shipped the *Gulfhawk II* to Europe where he performed before aviation enthusiasts in England, France, Holland, and Germany. During this overseas visit the only other person ever to fly the *Gulfhawk II,* the famous German World War I ace Ernst Udet, piloted the aircraft over Germany. In exchange, Williams became the first American to fly the vaunted Messerschmitt Bf 109 (see page 130). Interestingly, during this tour Williams also demonstrated the US Navy's and Marine Corps' successful dive-bombing techniques. Udet was so impressed that he incorporated these tactics into the Luftwaffe with devastating effect in World War II.

A new pilot's throat microphone was tried out in the *Gulfhawk II* for the first time in 1937, and during World War II the *Gulfhawk II* was used to test oils, fuels, and lubricants under extreme operating conditions. In 1942, at the request of Gen. Henry H. "Hap" Arnold, Williams made a three-month tour of fighter schools and training bases to demonstrate airmanship and precision aerobatic flying to pilot trainees.

In October 1948, *Gulfhawk II* was donated to the Smithsonian by the Gulf Oil Corporation. It was displayed in the Arts and Industries Building and in the National Air and Space Museum on the Mall, and it is now in the Steven F. Udvar-Hazy Center.

THE FIRST BRITISH monoplane fighter, the Hawker Hurricane, proved crucial during the Battle of Britain in the summer of 1940, and they continued to fight on nearly every front until the end of World War II.

The Hawker prototype was called the Interceptor Monoplane. It had retractable landing gear with an airframe of tubular metal cross-braced sections covered with fabric. While the prototype was under construction, the design was changed from a four-machine-gun armament to an armament of eight .303-caliber Browning weapons. Providing propulsion was a Rolls-Royce Merlin, twelve-cylinder, 990-hp engine driving a two-blade, fixed-pitch wooden propeller. It flew for the first time on November 6, 1935.

Engine troubles plagued the aircraft, and the Merlin engine was continually upgraded to increase performance and reliability. The Merlin III, capable of 1,030 hp, became the most satisfactory engine, and powered most of the early production aircraft.

As prototype evaluation continued, Hawker received an unprecedented contract order for 600 aircraft. In 1939, Hawker began using variable-pitch, three-bladed propellers,

2

3

DIMENSIONS: ——————

WINGSPAN	12.19 M (40 FT.)
LENGTH	9.6 M (31 FT., 5 IN.)
HEIGHT	4 M (13 FT., 1 IN.)
WEIGHT	2,630.88 KG (5,800 LB.)

1. The museum's Hawker Hurricane Mk.IIC on display at the Udvar-Hazy Center. **2.** A wartime photograph of a Hawker Hurricane Mk.IIC in flight over England. **3.** Royal Air Force pilots scrambling toward Hawker Hurricanes parked along the flight line sometime during the Battle of Britain, July 10 through October 31, 1940.

which increased the Hurricane's climb performance and service ceiling.

The Hurricane first saw combat after Hitler invaded Poland in 1939. Soon several Hurricane squadrons were heavily engaged with German aircraft above France, but the Luftwaffe outnumbered them and inflicted heavy casualties.

In the Battle of Britain fought between July and October 1940, twenty-six Hurricane I squadrons, with nineteen Supermarine Spitfire (see page 180) squadrons and ten other fighter squadrons equipped with obsolete Defiants and Blenheims—about 720 airplanes in all—fought against 2,000 Luftwaffe aircraft.

Their winning strategy called for the Hurricane squadrons to concentrate on the bombers while the faster Spitfires engaged the escorting fighters. While the German Messerschmitt Bf 109 (see page 130) outclassed the Hawker fighter in speed and armament, Hurricanes outturned the German fighter at all altitudes. Its easily repaired tubular construction also allowed many damaged machines to return quickly to the fight.

After the Battle of Britain many Hurricanes were modified to fly and fight in darkness and bad weather. Others were sent overseas to the Mediterranean. Hawker also sold Hurricanes to Yugoslavia, Belgium, Iran, Romania, Turkey, Poland, and the Soviet Union. Hurricanes also fought in Malta and the western desert of North Africa. The Hurricane even went to sea aboard CAM (Catapult Aircraft Merchantmen) ships. These hastily modified interceptors were nicknamed "Hurricats."

A larger Merlin XX engine that generated 1,280 hp and a maximum speed in the Hurricane of 550 km/h (342 mph) replaced earlier engines in the Hurricane II. Armament was increased from eight guns to twelve guns. More than 4,700 were built.

Several other variants, including the Mark IID with two 40-mm cannons mounted under the wings, were shipped to the Middle East, North Africa, and Burma, and one squadron flew them in northern Europe.

In the late 1960s the National Air and Space Museum obtained the Hawker Hurricane from the Royal Air Force Museum at Hendon. It was put on display after restoration that took eleven years to complete.

THE HUBBLE SPACE Telescope Structural Dynamic Test Vehicle in the National Air and Space Museum's Space Hall was originally created for conducting a wide range of engineering studies and later for simulating maintenance and repair activities in orbit.

The limitations on astronomy posed by the Earth's atmosphere have long been promoted as justification for placing telescopes in space. Two satellites with telescopes were put into space through the Orbiting Astronomical Observatory program in the mid-1960s and early 1970s. A successor telescope was to be launched on a Titan rocket, but NASA was converting to a shuttle-type launch vehicle. New specifications were made for a telescope that could fit into the space shuttle (see page 174) cargo bay. The final design of the Hubble Space Telescope, renamed after Edwin Hubble in 1983, contained at its core a 2.4-m (94-in.) Cassegrain-type telescope.

The main mirror is made from ultra-low expansion-fused silicate glass. The main and secondary mirrors are supported by a 4.9-m (192-in.) long graphite epoxy metering truss

DIMENSIONS: ═══════

LENGTH   12.9 M (42.3 FT.)

DIAMETER   3.27 M (10.7 FT.) AFT;

3.04 M (9.97 FT.) FORWARD

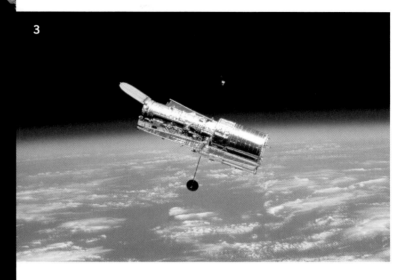

3

1. The museum's Hubble Space Telescope (HST) on display in the Space Race exhibition. 2. Attached to the "robot arm," the Hubble Space Telescope is unberthed and lifted up into the sunlight during this the second servicing mission on February 19, 1997. 3. Hubble Space Telescope (HST) in orbit after its deployment on this second servicing mission.

built to a tolerance of less than 0.002 cm. The telescope structure is encased in the cylindrical tube of the satellite. That tube is equipped with light baffles to keep stray light from reaching the main mirror. The tube terminates in an aperture door that closes off the tube when not in use. The support systems module that contains the various electronic and mechanical systems required to operate the observatory surrounds the lower end of the tube. The Hubble Space Telescope was designed to be serviced and upgraded in space.

Pointing of the telescope to a specific target is accomplished with three fine guidance sensors located in one of the four radial instrument bays behind the focal plane. They receive light from the stars selected for orientation from a 90-degree annulus ("pickle") at the edge of the field. Only two at one time are used for pointing, while the third provides data for mapping celestial objects.

In its original launch configuration Hubble included five additional instruments. The wide field/planetary camera occupied the remaining axial bay. The remaining instruments were located axially behind the focal plane. These consisted of the faint object camera, the faint object

spectrograph, the Goddard High Resolution Spectrograph, and the high-speed photometer. The Hubble Space Telescope was originally scheduled for launch in a shuttle mission in 1983, but was delayed. Then, the tragic *Challenger* explosion in 1986 led to a three-year hiatus in shuttle missions. The telescope was finally placed in orbit by *Discovery* in April 1990.

Spherical aberration was found in the Hubble's main mirror as the instrument was being calibrated. This flaw limited the resolution of images. A wealth of new astronomical data was obtained, however, in the first three years in spite of that. In late 1993 astronauts performed a major service and upgrade mission, which included the installation of new solar array panels and replacement of several guidance gyroscopes.

An optical correction system with a new set of specially configured mirrors to compensate for the main mirror's spherical aberration replaced one of the axial instruments. A new wide field/planetary camera was also installed with tiny corrections to its optics to compensate for the main mirror aberration. In February 1997 astronauts from the *Discovery* space shuttle performed a second service mission.

ON SEPTEMBER 13, 1935, Howard Hughes, noted pilot, millionaire businessman, and movie producer, flashed through the skies near Santa Ana, California, in his 1B Racer, breaking the world's absolute speed record for landplanes by flying 567 km/h (352.322 mph). This mark would stand unbroken for almost four years.

Impressed with the 1B, Hughes quickly decided to attempt to break the record for flying a transcontinental route across the United States in the same plane. The aircraft was refitted with longer wings and a new engine built for endurance. Hughes took off from Los Angeles on January 19, 1937, and 7 hours, 28 minutes, and 25 seconds later, landed at Newark Airport. His record-setting speed over the 3,862 km (2,400 mi.) was a remarkable 534 km/h (332 mph) and would have been higher had Hughes not been required to circle Newark Airport while waiting for a commercial airliner to land.

The Hughes 1B, also known as the H-1, was a milestone in aircraft design and pioneered the development of high-speed, high-powered, radial-engine aircraft, greatly influencing the subsequent design of a number of important single-engine fighters of World War II notoriety.

Designed by Dick Palmer, the 1B Racer was hand-built from wood and aluminum with careful attention to the smallest details. The fuselage was an aluminum semimonocoque with cantilevered tail surfaces. Flat, narrow aluminum sheets were butt-jointed together to form a smooth contour. Flush, countersunk rivets were used throughout, then shaved and polished to produce as clean a surface as possible. All fasteners and screwheads were aligned with the projected airflow. The cockpit was mounted behind the trailing edge of the wing and featured a Plexiglas windscreen that slid forward on tracks to allow entry. A

1. The museum's Hughes 1B on display in the Golden Age of Flight gallery. 2. Howard Hughes poses beside the tail of his Hughes 1B Racer. 3. Howard Hughes sitting in the cockpit of the Hughes 1B Racer. 4. The Hughes 1B Racer on the ground; note towing dolly under tail wheel.

DIMENSIONS: _____

WINGSPAN	9.67 M (31 FT., 9 IN.)
LENGTH	8.23 M (27 FT.)
HEIGHT	3.3 M (10 FT., 10 IN.)
WEIGHT	2,495 KG (5,500 LB.)

substantial keel provided excellent rigidity and strength to the airframe.

The wooden wing was built in one piece with two main box spars and numerous supporting ribs. The wing surface was formed by a plywood covering shaped to produce the optimum airfoil profile. Shaved and sanded, the wing was then covered in Wellington Balloon cloth that was then sealed, doped, and waxed to an extremely high gloss. The control surfaces were made of aluminum and covered in grade-A cotton fabric, doped in a deep royal blue to match the blue paint of the wing. The rudder and elevators were also made of aluminum and covered with grade-A cotton fabric and doped to match the aluminum fuselage.

The landing gear retracted hydraulically into the wing. A retractable tail skid was installed to assist in slowing the aircraft when landing.

A Pratt & Whitney SA1-G Twin Wasp Jr. engine rated at 700 hp was boosted to over 900 hp by resetting the carburetor and using 100-octane fuel. The engine was beautifully cowled and blended into the fuselage. A two-bladed Hamilton Standard controllable pitch propeller with external counterweights was fitted. Short exhaust stacks provided a small measure of additional thrust.

For the record-setting transcontinental flight, many changes were made. A new wing with greater span, area, and a different airfoil shape was installed, and the plane held up to a total of 1,060 l (280 gal.) of fuel.

A revised tail cone was fitted to compensate for the increased area of the tail surfaces. Additional navigational equipment as well as an oxygen system were also added. Hughes reengined his racer with a different Twin Wasp Jr. Taken from Jacqueline Cochran's Northrop Gamma, this powerplant was stressed in key areas for increased durability.

In February 1975 the 1B Racer was donated to the Smithsonian. Today it is part of the Golden Age of Flight gallery in the National Air and Space Museum.

THE GERMAN JUNKERS company, based at Dessau, Saxony, had pioneered the use of metal, specifically an aluminum alloy, when it introduced the Junkers F 13 in 1919.

During the 1920s, Junkers had already built several tri-motored large transports. Still using the same corrugated metal skin structural technology, designer Ernst Zindel produced the Junkers Ju 52, which made its first flight on September 11, 1930. Equipped with several large doors and a hatch in the roof, this was a single-engine aircraft intended for hauling freight. Its performance was impressive. In the winter of 1931, in Montreal, Canada, one took off carrying a load of almost four tons in 17.5 seconds. Due to the world's depressed economy, however, only seven Ju 52s were built.

In 1932 the Ju 52 was converted to a tri-motor, with three 525-hp BMW Hornet engines. Now designated Junkers Ju 52/3m, it carried up to seventeen passengers or about three tons of freight, and cruised at about 241 km/h (150 mph). Its best feature was its ability to take off from or land on almost any reasonably sized field, even a football field.

As an airliner, it was used all over Europe. The German Deutsche Luft Hansa (DLH) had more than 200 of them. Popular among pilots, it was affectionately known as *"Tante Ju"* ("Auntie Ju"). It was exported all over the world, serving in England, South America, China, and South Africa.

As a military transport, it was a great workhorse. Of the estimated 4,835 built, 2,804 were for the Luftwaffe, for which it performed valiantly during World War II as a troop carrier, bomber, and ambulance. Most spectacularly, an armada of Ju 52/3m's parachuted troops

2

DIMENSIONS:

WINGSPAN	29.25 M (95 FT., 11.5 IN.)
LENGTH	18.9 M (62 FT.)
HEIGHT	4.5 M (14 FT., 9 IN.)
WEIGHT	6,510 KG (14,354 LB.)

1. The museum's Lufthansa CASA 352L (license-built Junkers Ju 52/3m) on display at the Udvar-Hazy Center. **2.** A Syndicato Condor Junkers Ju 52/3m *Curupira* floatplane moored in the harbor at Rio de Janeiro, Brazil, circa mid-1930s. **3.** A Lufthansa Junkers Ju 52/3m *Erich Pust* in flight to Bathurst, West Africa, carrying air mail to South America in the mid-1930s. **4.** A posed photograph of the passenger cabin of a Deruluft Junkers Ju 52/3m.

into Allied-held Crete, and 170 of the fleet of 493 were shot down. Soviet sources claim that 676 were shot down or destroyed in the unsuccessful attempt to relieve von Paulus's army trapped in Stalingrad. Many of these flew to the battle zone fully loaded with supplies at the expense of the fuel needed to make the return flight.

Ju 52/3m's were produced in France during and after the war. In Spain, where it was designated CASA 352/3m, they were produced until 1952 and used extensively by the Spanish air force. The floatplane version of the Ju 52/3m was flown along the coast of Norway throughout World War II. Quite a few Ju 52/3m's continued flying after the war. The last one is believed to have been retired from commercial airline service in New Guinea during the late 1960s.

The Junker Ju 52/3m in the National Air and Space Museum was built in Spain. It was donated to the Smithsonian by Lufthansa in 1987.

1

IN FEBRUARY 1955 the high-level Killian Committee recommended to President Dwight D. Eisenhower that the United States develop both intercontinental ballistic missiles (ICBMs) and 2,414-km (1,500-mi.) intermediate-range ballistic missiles (IRBMs) as a matter of top priority. The president accepted the recommendation, and it was decided that the US Air Force build a ground-launched version, Thor, and the US Army and Navy together build a ship-launched version, Jupiter.

Developed under the direction of Dr. Werner von Braun, the liquid-fueled Jupiter IRBM began flight tests 1955, and the missile proved to be a reliable performer. One prototype flew 2,575 km (1,600 mi.) in May 1957, qualifying as the first successful IRBM launch by the United States. Within days of the launch of the Soviet Union's *Sputnik I* into orbit in October 1957, President Eisenhower ordered both the Thor and Jupiter into full production.

As finally produced, the Jupiter was 18 m (60 ft.) long with a diameter of 2.6 m (8 ft., 9 in.). It weighed over 48,988 kg (108,000 lb.) fully fueled and had a single S-3D engine

1. A Jupiter missile in a nighttime view. **2.** The museum's Jupiter C rocket in the center on display in the Space Race hall. **3.** Monkey Baker perches on a model of the Jupiter intermediate-range ballistic missile (IRBM) that launched her into space on May 28, 1959. **4.** View of the Jupiter C launch vehicle in a gantry at the Redstone Arsenal, Huntsville, Alabama, in June 1958.

that used liquid oxygen. The missile had a range of 2,414 km (1,500 mi.) and carried one W-49 nuclear warhead with a yield of approximately 1.44 megatons. Chrysler Corporation's Ballistic Missile Division produced the airframes, North American Aviation's Rocketdyne Division the engine, Ford Instrument Company the guidance system, and General Electric the reentry vehicle.

The State Department negotiated with various NATO allies to accept the Jupiters. In April 1958 it appeared that France had agreed to base three squadrons (45 missiles). However, last-minute problems arose, and the United States had to look elsewhere. Italy agreed to accept Jupiters in late 1958 and Turkey in late 1959. Two squadrons were deployed to Italy and the first missiles became operational in July 1960. One squadron was deployed to Turkey, and the first missiles there became operational in 1962.

The Jupiters added little to the strategic power of the United States for several reasons. First, they were based above ground in fixed positions and took several hours to get ready for launch, making them vulnerable to attack and destruction. Second, by the time they became operational the United States would have started to deploy significant numbers of Polaris submarine-launched ballistic missiles (SLBMs) and Atlas, Titan I, and Minuteman ICBMs. These missiles could deliver many more nuclear weapons and were considerably less vulnerable to attack.

In May 1959 a Jupiter carried two monkeys, Able and Baker, 483 km (300 mi.) high and 2,575 km (1,600 mi.) downrange, and they were recovered alive. They were the first living payloads successfully launched and recovered alive by the United States. A Juno I, made from a modified Jupiter C, successfully launched the first American satellite, *Explorer 1,* in 1957. An extended Jupiter served as the booster stage for the Juno II, a rocket used by NASA to launch satellites from 1958 to 1961.

The Jupiter model in the National Air and Space Museum was acquired from NASA's Marshall Space Flight Center in 1971.

IN SEPTEMBER 1940, the Japanese Navy issued a specification for floatplane fighters capable of supporting offensive naval operations. Tests showed that the speed of the new Kawanishi N1K1 Kyufu (REX) was only slightly less than the Mitsubishi A6M Zero (see NASM collection), and the amphibious fighter was almost as maneuverable as its land-based cousin. This was remarkable performance for an aircraft that could not retract or jettison its huge landing gear.

Long before the first Kyofu flew, Kawanishi engineers believed that the basic design would also make an excellent land-based fighter. The conversion entailed replacing the main and wingtip floats with a conventional landing gear, fitting a more powerful and larger 2,000 horsepower engine and modifying the wings. Kawanishi flew the first N1K1-J Shiden (Violet Lightning) land-based fighter on December 27, 1942. Code named "George" by the Allies, the new airplane fell short of its projected speed of 649 kph (403 mph). Nevertheless, at 575 kph (357 mph), the N1K1-J was faster than the Mitsubishi A6M Zero ZEKE, and the Japanese Navy badly needed an effective counter to new American naval fighter aircraft such as the Grumman F6F Hellcat and Vought F4U Corsair. The Japanese Navy ordered Kawanishi to abandon two other fighter projects and start building Shidens.

By the end of 1943, Kawanishi delivered about seventy of the new fighters, which the navy used for pilot familiarization and training. Expecting Allied amphibious landings in the Philippines, the navy sent the first George unit to Cebu in October 1944. Even with engine, landing gear, logistics, and maintenance problems, the George units fought well against the Allied pilots.

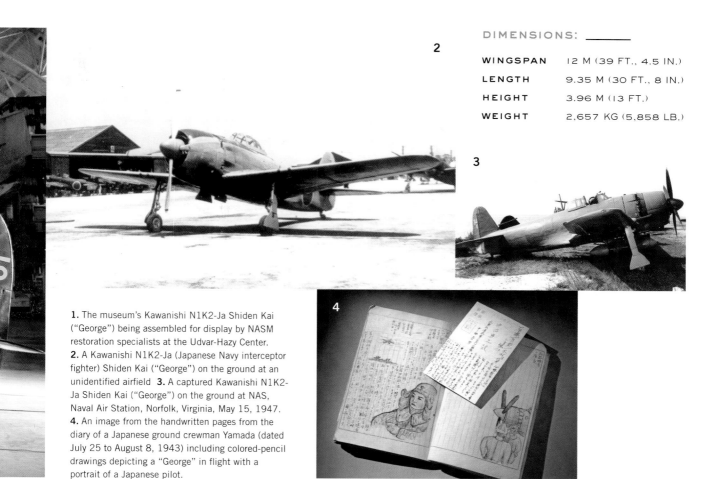

DIMENSIONS: _____

WINGSPAN	12 M (39 FT., 4.5 IN.)
LENGTH	9.35 M (30 FT., 8 IN.)
HEIGHT	3.96 M (13 FT.)
WEIGHT	2,657 KG (5,858 LB.)

1. The museum's Kawanishi N1K2-Ja Shiden Kai ("George") being assembled for display by NASM restoration specialists at the Udvar-Hazy Center. **2.** A Kawanishi N1K2-Ja (Japanese Navy interceptor fighter) Shiden Kai ("George") on the ground at an unidentified airfield **3.** A captured Kawanishi N1K2-Ja Shiden Kai ("George") on the ground at NAS, Naval Air Station, Norfolk, Virginia, May 15, 1947. **4.** An image from the handwritten pages from the diary of a Japanese ground crewman Yamada (dated July 25 to August 8, 1943) including colored-pencil drawings depicting a "George" in flight with a portrait of a Japanese pilot.

Kawanishi set about refining the design. They lowered the wings from mid-fuselage, and the extra ground clearance permitted the engineers to install a shorter, more conventional and less troublesome landing gear, simplify the fuselage structure, and redesign the empennage. Only the wings and armament remained from the initial design. The engine remained problematic, but the navy was impressed with these improvements and ordered the new version into production as the N1K2-Ja Shiden Kai.

In air-to-air combat, experienced Japanese pilots flying Shiden Kais more than held their own against most American pilots flying F6F Hellcats. Flying intercept missions against Boeing B-29 Superfortresses (see page 28) above the home islands, the Shiden Kai was not successful because of inadequate climb speed and power loss at high altitudes.

Kawanishi developed several other variants and had more planned when the war ended. About 1,500 of the various models were produced. In battle over Formosa (Taiwan), the Philippines, Okinawa, and the home islands, Shiden pilots acquitted themselves well, but this excellent airplane was another good design that appeared too late and in too few numbers to reverse Japan's fortunes in the air war.

The National Air and Space Museum's Kawanishi N1K2-Ja Shiden Kai is one of three remaining today. It was acquired from the US Navy in 1983. After three years of restoration, the Shiden Kai wears the colors and markings of the 343rd Kokutai, a unit stationed at Omura Naval Air Station in 1945.

SAMUEL PIERPONT LANGLEY (1834–1906) was a leading scientific figure in the United States in the late nineteenth century, well known especially for his astronomical research. He became the third secretary of the Smithsonian Institution in 1887.

Langley had begun serious investigation of theoretical aerodynamics several years earlier and began working on an actual flying machine in 1891. After several failures, Langley had his first genuine success on May 6, 1896, when the unpiloted Aerodrome No. 5 made the first successful flight. It was launched from a spring-actuated catapult mounted on top of a houseboat on the Potomac River near Quantico, Virginia. Two flights were made that afternoon, one of 1,005 m (3,300 ft.) and a second of 700 m (2,300 ft.), at a speed of approximately 40 km/h (25 mph). On November 28, another successful flight was made with a similar model, the Aerodrome No. 6, that flew approximately 1,460 m (4,790 ft.).

Serious work on the airplane, referred to as the Great Aerodrome or Aerodrome A, began in October 1898 after Langley received a $50,000 grant from the War Department.

The Aerodrome A was based on the successful performance of a gasoline-powered model that was one-fourth the size. This exact-scale miniature, known as the Quarter-Scale Aerodrome, flew satisfactorily twice on June 18, 1901, and again with an improved engine on August 8, 1903. But these successes masked its flaws as a design prototype for the full-sized, piloted airplane.

Charles M. Manly, Langley's assistant, created a water-cooled radial engine that generated a remarkable 52.4 hp at 950 rpm with a power-to-weight ratio of 1.8 kg (4 lb.) per horsepower, an amazing achievement for the time.

DIMENSIONS:

WINGSPAN	14.8 M (48 FT., 5 IN.)
LENGTH	16.0 M (52 FT., 5 IN.)
HEIGHT	3.5 M (11 FT., 4 IN.)
WEIGHT	340 KG (750 LB.)

1. The Langley Aerodrome A hanging on display at the museum's Udvar-Hazy Center. **2.** Charles Manly (left) pilot and Dr. Samuel Pierpont Langley (right) inventor and builder of the Langley Aerodrome A aircraft. **3.** The Langley Aerodrome A on its launching catapult onboard Dr. Samuel Pierpont Langley's houseboat on the Potomac River; circa October 7, 1903.

The airframe was an entirely different matter. It was structurally weak and unsound. Like the smaller aerodromes, it was a tandem-winged design with a cruciform tail. The control system was minimal.

For propulsion, two pusher propellers mounted between the tandem wings were driven by shafts and gears connected to the centrally mounted engine, again after the pattern of the smaller aerodromes. The huge aircraft spanned nearly 15 m (50 ft.) and was more than 16 m (52 ft.) long. It weighed 340 kg (750 lb.) with the pilot, Manly.

The first test flight of the Aerodrome A was on October 7, 1903. The airplane was assembled on the rear of a catapult track mounted on a large houseboat moored near Widewater, Virginia, close to the site where the small aerodromes had successfully flown. Immediately after launching, the Aerodrome plunged into the river at a 45-degree angle. Langley was bitterly disappointed and blamed the launch mechanism, not the aircraft.

After repairs, a second attempt was made on December 8, 1903, from the houseboat on the Potomac

River in Washington, DC. The results were equally disastrous, momentarily trapping Manly underneath the wreckage in the freezing Potomac. Langley again blamed the launching device.

While the catapult likely contributed to the failure, the Aerodrome A was an overly complex, structurally weak, and aerodynamically unsound aircraft. This second crash ended the aeronautical work of Samuel Langley. His request for further funding was refused, and he suffered much public ridicule.

The remains of the Aerodrome A were left with the Smithsonian by the War Department. Smithsonian officials at first misleadingly identified the Aerodrome A in its label text as the world's first airplane "capable of sustained free flight." This action was, partly, what prompted Orville Wright in 1928 to lend the 1903 Flyer (see page 194) to the Science Museum in London as a gesture of protest until the Wright brothers were given full credit for having invented the airplane. The Aerodrome A is on public view now in the Steven F. Udvar-Hazy Center.

IN 1959, WILLIAM P. LEAR, SR. initiated the development of the Lear Jet 23, the founding product of the Lear Jet Corporation, a pioneer in the new field of business and personal jet aviation. An example of aircraft form and function, the Lear Jet 23's ability to cruise above most weather, over long distances, and at high speed made it an outstanding performer.

Based on the AFA P-16, a Swiss strike-fighter, performance of the Lear Jet 23 was the direct result of excellent design and construction. The fuselage narrows at each side as the wing and engine nacelles extend outward. This is a design concept known as "area rule" and is used to smooth the flow of air around these projections during high-speed flight.

The cabin seats up to nine people, including pilots, and may be fully pressurized. The windshield and cabin windows are formed of stretched and laminated acrylic plastic. The ailerons, elevators, and rudder are mechanically connected to the cockpit controls, while all trim surfaces are electrically operated. There is a small tab on the left aileron for roll and one on the rudder for yaw. The incidence angle of the horizontal stabilizer is changed for pitch trim control. The 23 has spoilers for speed control on the upper surface of each wing panel forward of the flaps. These are actuated by hydraulic pressure, as is the fully retractable landing gear and multiple disc brakes on the dual mail wheels.

On October 7, 1963, prototype Model 23 made its first flight. This plane flew 194 hours during 167 separate flights until it was destroyed. On May 21, 1965, pilots John M. Conroy and Clay Lacey covered 3,110 km (5,005 mi.) from Los Angeles to New York and back in 11 hours, 36 minutes, in a Lear Jet 23, setting a record. On December 14, 1965,

DIMENSIONS: ═══

WINGSPAN	10.8 M (35 FT., 7 IN.)
LENGTH	13.2 M (43 FT., 3 IN.)
HEIGHT	3.8 M (12 FT., 7 IN.)
WEIGHT	2,790 KG (12,750 LB.)

1. The museum's Lear Jet 23 on exhibit at the Udvar-Hazy Center, January 9, 2004. **2.** The Lear Jet 23 prototype in flight. The aircraft retains the nose probe for flight test instrumentation. **3.** William Powell "Bill" Lear Sr., chairman of the board of Lear, Inc. **4.** NASM's Lear Jet 23 about to touch down at an unidentified airport.

another Model 23 reached an altitude of 12.1 km (40,000 ft.) in 7 minutes, 21 seconds. With an engine thrust-to-weight ratio of 1 to 2.2 pounds, a Model 23 can outclimb an F-100 Super Sabre to 3 km (10,000 ft.), and can be just as impressive on the way down.

The success of the basic Model 23 design and the expansion of the corporate and personal jet market inspired a number of derivative models with increased range, size, and speed. A Model 24 flew around the world and set or broke eighteen international aviation records.

And Neil Armstrong set five world records for an aircraft of this class: two for altitude achieved, two for sustained flight at 15.5 km (51,000 ft.), and one for high-altitude time-to-climb.

The Lear Jet 23 in the National Air and Space Museum is the second one built. After 1,127 flying hours, the aircraft was returned to the factory for restoration, and the Gates Learjet Corporation donated it to the Smithsonian in 1977.

THE MOST SIGNIFICANT pre–Wright brothers aeronautical experimenter was the German glider pioneer Otto Lilienthal.

He began research in aeronautics with his brother, Gustave, in the late 1860s, investigating the mechanics and aerodynamics of bird flight. In the 1870s he conducted a series of experiments on wing shapes and gathered air pressure data using a whirling arm in the natural wind. The research produced the best and most complete body of aerodynamic data of the day.

Lilienthal established definitively the widely held belief that a curved wing section, as opposed to a flat wing surface, was the optimum shape for generating lift. He published his findings in a pathbreaking book called *Der Vogelflug als Grundlage der Fliegekunst* (*Birdflight as the Basis of Aviation*) in 1889.

Between 1891 and 1896, he put his research into practice in the form of a series of highly successful full-scale glider trials. During this period, Lilienthal made close to 2,000 brief flights in sixteen different glider designs. Most of these were monoplanes with stabilizing tail surfaces mounted at the rear, but he also tried a few biplane and folding-wing designs; the original monoplane glider, or *Normal-Segelapparat* (standard sailing machine), as he called it, produced the best results.

The gliders had split willow frames covered with cotton-twill fabric sealed with collodion to make the surface as airtight as possible. Collodion is a viscous solution of nitrated cellulose in a mixture of alcohol and ether that dries to form a tough elastic film. The wings ranged in area from 9 to 25 sq. m (100 to 280 sq. ft.) and could be folded to the rear for transport and storage. Control was achieved by shifting body weight, similar to

2

DIMENSIONS: ⎯⎯⎯⎯

WINGSPAN	7.93 M (26 FT.)
LENGTH	4.19 M (13 FT., .75 IN.)
HEIGHT	1.53 M (5 FT.)

3

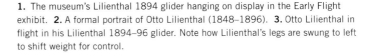

1. The museum's Lilienthal 1894 glider hanging on display in the Early Flight exhibit. **2.** A formal portrait of Otto Lilienthal (1848–1896). **3.** Otto Lilienthal in flight in his Lilienthal 1894–96 glider. Note how Lilienthal's legs are swung to left to shift weight for control.

modern hang gliders. Suspended in a harness and swinging his legs from side to side and fore and aft, the pilot could adjust the center of gravity and thereby maintain equilibrium.

Lilienthal's best efforts with these gliders covered more than 300 m (985 ft.) and were 12 to 15 seconds in duration. On August 9, 1896, Lilienthal's aeronautical experiments came to an abrupt and tragic end while he was soaring in one of his standard monoplane gliders. A strong gust of wind caused the craft to nose up sharply, stall, and crash from an altitude of 15 m (50 ft.). Lilienthal suffered a broken spine and died the following day in a Berlin hospital.

As successful as they were, Lilienthal's glider designs had some inherent limitations that he would have had to confront had he lived. The principal problem was his means of controlling the craft. Lilienthal's technique of shifting body weight to maintain equilibrium did place him ahead of others in that he recognized the need for a control

system, but, as revealed in his fatal crash, the control response of his method was limited.

Despite unresolved issues, the impact of Lilienthal's aeronautical work on the next generation of experimenters was great. Lilienthal established a new starting point for anyone entering the field. Beyond his technical contributions, he sparked aeronautical advancement from a psychological point of view as well. He demonstrated unquestionably that gliding flight was possible.

Lilienthal's trips through the air made headlines everywhere, and dramatic photographs showing him soaring gracefully appeared in newspapers and magazines the world over. The publicity made him quite a sensation in an age when, for most, human flight still seemed a distant possibility.

The Lilienthal glider in the National Air and Space Museum, one of only six remaining in the world, was built by the German experimenter in 1894. It was presented to the Smithsonian in 1906.

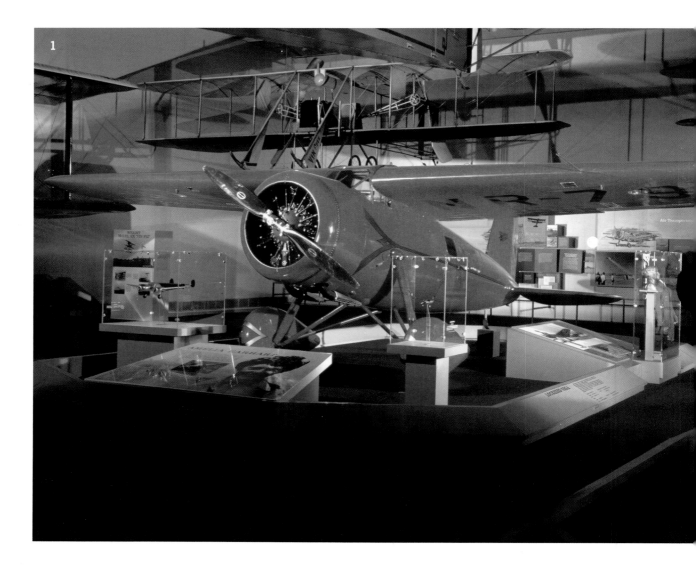

IN THE BRIGHT red Lockheed 5B Vega now on display in the National Air and Space Museum, Amelia Earhart became the first woman, and only the second person, to fly solo nonstop across the Atlantic Ocean and the first woman to fly solo nonstop across the United States.

Earhart wanted her record-breaking flight to be in a Lockheed Vega, a sleek, new monoplane. Used by Lockheed as a demonstrator, Earhart bought it in March 1930. During a test flight the plane nosed over on landing, causing damage to the fuselage, which took over a year to repair. During the repair, and for the ocean crossing, the plane was also strengthened to accommodate the extra fuel tanks added to provide a 92-l (420-gal.) capacity. Additional instruments were also installed.

On May 20, 1932, Earhart set off alone from Harbor Grace, Newfoundland. The weather was a problem from the start, and at one point in the flight ice on the wings forced her into a 914-m (3,000-ft.) unchecked descent. She managed to level off and, constantly fighting fatigue, she landed in a field near Culmore, Londonderry, Northern Ireland. She had made the 3,256-km (2,026-mi.) flight in about 15 hours.

Earhart was never content to rest on her laurels. On August 24, 1932, in the same Vega, she flew from Los Angeles to Newark, covering the 3,940 km (2,448 mi.) in 19 hours, 5 minutes. This was the first solo nonstop transcontinental flight by a woman.

WINGSPAN	12.49 M (41 FT.)
LENGTH	8.38 M (27 FT., 6 IN.)
HEIGHT	2.49 M (8 FT., 2 IN.)
WEIGHT	748 KG (1,650 LB.)

1. Amelia Earhart's Lockheed 5B Vega on display in the museum's Pioneers of Flight gallery. 2. Amelia Earhart standing on steps beside her Lockheed 5B Vega in Culmore, North Ireland, after her historic solo flight across the Atlantic, May 21, 1932. 3. Amelia Earhart, Allen Loughead, Carl Squire, and Lloyd Stearman stand in front of Earhart's Lockheed 5B Vega at the Lockheed hangar in Burbank, California. 4. The cover of the banquet program with menu from the National Aeronautic Association of the U.S.A., Paris chapter, dinner in honor of Amelia Earhart. Paris, June 3, 1932.

Amelia Earhart's 5B was one of 131 Vegas manufactured. It was an aesthetically pleasing aircraft with a spruce veneer monocoque fuselage and a spruce cantilever wing. In an era of biplanes, the 5B featured an advanced monoplane configuration in which the wing was completely braced internally.

The Vega reintroduced the monocoque fuselage shape, which had first appeared in racing airplanes around 1913. This shape, wherein the fuselage is essentially a shell, maximizes both the load-carrying ability of the aircraft and its useful internal space. It subsequently became a standard design practice for transport aircraft.

The Smithsonian proudly displays Earhart's Vega in the National Air and Space Museum's Pioneers of Flight gallery. It was donated by the Franklin Institute in 1966.

THE NATIONAL AIR and Space Museum is fortunate to have as part of its collection *Winnie Mae,* a special Lockheed 5C Vega flown by famed aviator Wiley Post twice around the world and in a series of important substratospheric research flights.

Purchased in June 1930, it is named for the daughter of its original owner, F. C. Hall, who hired Post to pilot the plane. Post entered *Winnie Mae* in the National Air Races and piloted the plane to the first of its records. Inscribed on the side of its fuselage is "Los Angeles to Chicago 9 hrs. 9 min. 4 sec. Aug. 27, 1930."

On June 23, 1931, Post, accompanied by Harold Gatty as navigator, took off from New York to make the world circuit. The first stop was Harbour Grace, Newfoundland. From there, the fourteen-stop course included England, Germany, Russia, Siberia, Alaska, Canada, and Cleveland. The circuit was completed in 8 days, 15 hours, and 51 minutes. Hall's admiration for his pilot manifested itself in the gift of the *Winnie Mae* to Post.

Wiley Post spent the following year exhibiting the plane and conducting various flight tests. The airplane was given an overhaul of the engine, a radio compass, and an autopilot. Both the instruments were in their final stages of development by the army and Sperry Gyroscope Company.

Leaving New York on July 15, 1933, and closely following his former route around the world, this time making only eleven stops, Post made the 25,099-km (15,596-mi.) trip in 7 days, 18 hours, and 49 minutes.

Post next modified *Winnie Mae* for long-distance, high-altitude operation. Needing to

DIMENSIONS: ════

WINGSPAN	12.49 M (41 FT.)
LENGTH	8.38 M (27 FT., 6 IN.)
HEIGHT	2.49 M (8 FT., 2 IN.)
WEIGHT	1,177 KG (2,595 LB.)

1. Wiley Post's Lockheed Vega 5C *Winnie Mae* on display at the Udvar-Hazy Center. **2.** Wiley Post stands beside his Lockheed 5B Vega *Winnie Mae* at Floyd Bennett Field, Long Island, New York, June 19, 1933. **3.** Wiley Post in front of his Lockheed 5 Vega *Winnie Mae* in or around Bartlesville, Oklahoma in the early 1930s. **4.** The first high-altitude pressure suit worn by Wiley Post in 1935.

develop some means of enabling the pilot to operate in a cabin atmosphere of greater density than the outside atmospheric environment, Post asked the B. F. Goodrich Company to assist him in developing a full pressure suit. Post hoped that by equipping the plane with an engine supercharger and special jettisonable landing gear, and himself with a pressure suit, he could cruise for long distances at high altitude in the jet stream. On March 15, 1935, Post flew from Burbank, California, to Cleveland, Ohio, a distance of 3,275 km (2,035 mi.), in 7 hours and 19 minutes. At times, the *Winnie Mae* attained a ground speed of 547 km/h (340 mph), indicating that the airplane had been in the jet stream.

Wiley Post died shortly afterward in the crash of a hybrid Lockheed Orion-Sirius floatplane. His companion, humorist Will Rogers, also perished in the accident. The Smithsonian acquired the *Winnie Mae* from Mrs. Post in 1936.

During its high-altitude flight research, the *Winnie Mae* made use of a special tubular steel landing gear developed by Lockheed engineers Clarence L. "Kelly" Johnson and James Gerschler that was released after takeoff by the pilot using a cockpit lever, thus reducing the total weight and drag of the plane. The *Winnie Mae* would then land on a special metal-covered spruce skid glued to the fuselage.

During these flights, Post wore the world's first practical pressure suit. Developed by Post and Russell S. Colley of B. F. Goodrich Company, it consisted of three layers: long underwear, an inner black rubber air pressure bladder, and an outer cloth contoured suit. A special pressure helmet was then bolted on the suit that had a removable faceplate that Post could seal when he reached a height of 5,181.6 m (17,000 ft.). The helmet had a special breathing oxygen system and could accommodate earphones and a throat microphone.

Winnie Mae, its special jettisonable landing gear, and Post's pressure suit are part of the collection of the National Air and Space Museum.

A VACATION FLIGHT with "no start or finish, no diplomatic or commercial significance, and no records to be sought," is what Charles A. Lindbergh said about the flight that he and his wife, Anne Morrow Lindbergh, were planning to make to the Orient in 1931. Their choice of route, however, showed the feasibility of using the great circle to reach the Far East.

The Lindberghs flew in a Lockheed Sirius low-wing monoplane, powered by a 680-hp Wright Cyclone engine. Their airplane was specially fitted with Edo floats, because most of the flight was to be over water.

Their route took them from North Haven, Maine, to Ottawa, Moose Factory, Churchill, Baker Lake, and Aklavik, all in Canada; Point Barrow, Shismaref, and Nome, Alaska; Petropavlosk, Siberia; and on over the Kurile Islands to Japan. After receiving an enthusiastic welcome in Tokyo, they flew to China. They landed on Lotus Lake near Nanking on September 19, completing the first flight from the West to the East by way of the North.

At Hankow, the Sirius, with the Lindberghs aboard, was being lowered into the Yangtze River from the British aircraft carrier *Hermes* when it accidentally capsized. One of the wings was damaged when it hit one of the ship's cables. The Sirius had to be returned to the United States for repairs.

In 1933, Pan American Airways, Imperial Airways of Great Britain, Lufthansa of Germany, KLM of Holland, and Air France undertook a cooperative study of possible Atlantic routes. Each was assigned the responsibility for one of the following areas: Newfoundland to Europe via Greenland; Newfoundland via the great circle route to Ireland; Newfoundland

2 **3**

4

1. Charles Lindbergh's Lockheed Sirius 8 *Tingmissartoq* display in the Pioneers of Flight gallery at the National Air and Space Museum. **2.** Anne and Charles Lindbergh pose in front of their Lockheed Sirius at the beaching platform of the EDO Corporation, College Point, Long Island, New York. **3.** The *Tingmissartoq* on the water at Angmassalik, Greenland, 1933. An Angmassalik boy has just started to paint the nickname *Tingmissartoq* on the side of the fuselage. **4.** A view of Lindbergh on the left wing of his Lockheed Sirius, inverted on hoist, in Yangtse River, after handling accident during launch on October 2, 1931.

DIMENSIONS: ————

WINGSPAN	13.05 M (42 FT., 10 IN.)
LENGTH	9.14 M (30 FT.)
HEIGHT	4.5 M (14 FT., 9 IN.)
WEIGHT	2,082 KG (4,589 LB.)

southeast to the Azores and Lisbon; Miami, Bermuda, the Azores, and Lisbon; and across the South Atlantic from Natal, Brazil, to Cape Verde, Africa.

Ground-survey and weather crews in Greenland were already hard at work when Lindbergh, Pan Am's technical advisor, took off from New York on July 9 in the rebuilt Lockheed Sirius, again accompanied by his wife, who served as copilot and radio operator. An artificial horizon and a directional gyro had been added to the instrument panel since the previous flight, and a new Wright Cyclone SR1820-F2 engine of 710 hp was installed. Lindbergh's plan was not to set up a particular route but to gather as much information as possible on the area to be covered.

Every possible space in the aircraft was utilized, including the wings and floats, which contained the gasoline tanks. There was plenty of emergency equipment in case the Lindberghs had to make a forced landing in the frozen wilderness.

From New York, the Lindberghs flew up the eastern border of Canada to Hopedale, Labrador. From Hopedale they made the first major overwater hop, 1,046 km (650 mi.) to Godthaab, Greenland, where the Sirius acquired the name *Tingmissartoq,* which in Inuit means "one who flies like a big bird."

After crisscrossing Greenland to Baffin Island and back, and then on to Iceland, the Lindberghs proceeded to the major cities of Europe and as far east as Moscow, down the west coast of Africa, and across the South Atlantic to South America, where they flew down the Amazon, and then north through Trinidad and Barbados and back to the United States.

They returned to New York on December 19, having traveled 48,280 km (30,000 mi.) to four continents and twenty-one countries. The information gained from the trip proved invaluable in planning commercial air transport routes for the North and South Atlantic.

The Lindberghs' 8 Sirius was transferred to the Smithsonian from the American Museum of Natural History in 1959.

THE PREDECESSOR to the Lockheed SR-71A, and designated A-11 by the Central Intelligence Agency, the Lockheed SR-71A Blackbird was designed to cruise at Mach 3.2 and fly well above 18,288 m (60,000 ft.).

Flying more than three times the speed of sound generates 316°C (600°F) temperatures on external aircraft surfaces, which is enough to melt conventional aluminum airframes. Because of this, the A-11 was covered with an external skin of titanium alloy. Two powerful afterburning turbine engines propelled this remarkable aircraft from a takeoff speed of 334 km/h (207 mph) to more than 3,540 km/h (2,200 mph). To prevent supersonic shock waves from causing flameouts, Clarence "Kelly" Johnson's team designed a complex air intake and bypass system for the engines.

By carefully shaping the airframe to reflect as little transmitted radar energy as possible and covering it with a paint that absorbs radar waves, the A-11 was supposed to be invisible to radar detection. This was one of the earliest applications of stealth technology to aircraft.

After technical refinement, the CIA flew their first operational A-11 sortie on May 31, 1967, over North Vietnam. The US Air Force ordered an interceptor version of the aircraft

DIMENSIONS:

WINGSPAN	23.71 M (77 FT., 10 IN.)
LENGTH	15.18 M (49 FT., 10 IN.)
HEIGHT	4.16 M (13 FT., 8 IN.)
WEIGHT	3,470 KG (7,650 LB.)

1. The museum's Lockheed SR-71A Blackbird on display at the Udvar-Hazy Center. **2.** A beautifully lit view of the museum's SR-71A in its storage building at the Dulles International Airport. **3.** A Lockheed SR-71A Blackbird undergoing aerial refueling from Boeing KC-135 tanker.

designated the YF-12A. Configured to conduct post-nuclear-strike reconnaissance, this plane became the SR-71 and featured greater fuel capacity, a longer fuselage, and aerodynamic "chines" extended to the nose.

The first of thirty-two SR-71s flew on December 22, 1964. Because of extreme operational costs, military strategists decided that the more capable US Air Force SR-71s should replace the CIA's A-11s, which were retired in 1968. The US Air Force's 1st Strategic Reconnaissance Squadron took over the missions beginning in the spring of 1968, and it acquired the unofficial name Blackbird for the special black paint on the body.

Flying the SR-71 safely required two crew members, a pilot and a reconnaissance systems officer (RSO) who operated the aircraft sensors and defensive systems. In addition to an array of high-resolution cameras, the aircraft also carried equipment to record the strength, frequency, and wavelength of signals emitted by radars.

The SR-71 was designed to fly deep into hostile territory and avoid interception with its tremendous speed and high altitude. It could operate safely at a maximum

speed of Mach 3.3 at an altitude more than 25,908 m (85,000 ft.) above the Earth. The crew had to wear pressure suits similar to those worn by astronauts.

US Air Force pilots flew the SR-71 from Kadena Air Base, Japan, throughout its operational career but other bases hosted Blackbird operations, too. The 9th Strategic Reconnaissance Wing occasionally deployed from Beale Air Force Base in California to other locations to carry out operational missions.

Because the performance of orbiting reconnaissance satellites grew, along with the effectiveness of ground-based air defense networks, the USAF ceased SR-71 operations in 1995. NASA retained two SR-71As and one SR-71B for research projects until 1999.

On March 6, 1990, the service career of one Lockheed SR-71A Blackbird ended with a record-setting flight. Lt. Col. Ed Yielding and his RSO, Lt. Col. Joseph Vida, flew this aircraft from Los Angeles to Washington, DC, in 1 hour, 4 minutes, and 20 seconds, averaging a speed of 3,418 km/h (2,124 mph). The SR-71A was then transferred by the USAF to the National Air and Space Museum.

STILL SHROUDED IN secrecy thirty-five years after its creation, the Lockheed U-2 was one of the most successful intelligence-gathering aircraft ever produced.

In 1953, on behalf of the Central Intelligence Agency, the US Air Force issued a request for a single-seat, long-range, high-altitude reconnaissance aircraft to monitor the military activities of the Soviet Union and its satellite countries in Eastern Europe. By this time, breakthroughs in film and camera technologies made it possible to take high-resolution photographs of strategic sites from extreme altitudes where an airplane was thought invulnerable to interception.

On July 4, 1956, a U-2A completed the first flight over the Soviet Union. Sophisticated electronic and camera equipment was housed in the nose and in a large fuselage bay. Large fuel tanks enabled the aircraft to fly for six hours at transcontinental ranges at altitudes in excess of 18,288 m (60,000 ft.).

Operational U-2As flew routinely from bases in Pakistan and Turkey to Norway, over vast stretches of the Soviet Union. These flights gathered much important data and revealed that the so-called "missile gap" in the Soviets' favor was a myth.

For four years the CIA flew U-2As and improved U-2Bs until May 1, 1960, when Francis Gary Powers was shot down by a Soviet SA-2 missile over Sverdlovsk (now Ekaterinburg), sparking an embarrassing diplomatic incident for the United States that halted these flights.

Flights over the People's Republic of China, however, continued unabated from bases

DIMENSIONS:

WINGSPAN	24.4 M (80 FT.)
LENGTH	15.2 M (50 FT.)
HEIGHT	4.6 M (15 FT.)
WEIGHT	5,929 KG (13,071 LB.)

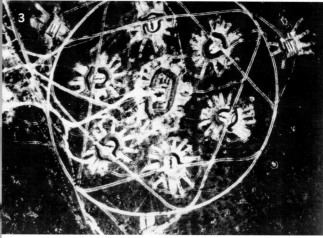

1. The museum's Lockheed U-2C on display in the Looking at Earth gallery. **2.** A Lockheed U-2A in flight. The US Air Force markings are visible on rear and forward fuselage and wings. **3.** A color aerial reconnaissance photograph taken by a U-2 showing a Soviet SA-2 surface-to-air missile (SAM) site in La Coloma, Cuba, November 10, 1962, during the Cuban Missile Crisis. **4.** A portrait of Francis Gary Powers standing with Lockheed U-2B. Powers is wearing an MC-2 high-altitude partial pressure suit and MA-1 helmet.

in Taiwan. Midway through 1962, a U-2 confirmed the presence of Soviet intermediate-range ballistic missiles in Cuba, which led to the Cuban Missile Crisis. U-2s were also in demand in Vietnam after July 1964, operating continually until the fall of Saigon in 1975. Since then, U-2s have observed the developing situations around the globe.

NASA has operated two U-2s in its High Altitude Missions Branch for stratospheric sampling, such as gathering volcanic dust after the 1980 eruption of Mount St. Helens. They have also been involved in assessments of natural disasters and water and land use.

Numerous versions of the U-2 have been produced, including two-seat models and models that can be operated from aircraft carriers. The most significant modification was for the U-2R, a redesign that lengthened both the fuselage and wingspan, allowing for much improved handling and a greater electronics and sensor payload.

On August 30, 1982, the National Air and Space Museum acquired its Lockheed U-2C from the US Air Force. The paint scheme on the aircraft is that of one used by the air force during operations from British bases in the Middle East.

THE LUNAR MODULE 2 (LM2) is unique among objects in the National Air and Space Museum in that it is the only true spaceship designed to operate solely in the vacuum of space.

Grumman Aerospace built fourteen lunar modules (LM) for NASA. Modules 1 through 9 were all nearly identical to one another. Numbers 10 through 14 were called the "extended stay" lunar modules because they were modified for the Apollo astronauts to spend several days on the Moon and to carry a great deal of equipment—including the lunar rover.

The first two modules were assigned to unmanned flights to test their performance in space. However, the flight of LM1 onboard *Apollo 5* was so successful that a second flight was considered unnecessary. Instead, LM2 was used in drop tests to ensure proper performance of the landing gear. LM2 is now in the National Air and Space Museum.

DIMENSIONS: ————

HEIGHT	7 M (23 FT.)
WIDTH	9.4 M (31 FT.)
WEIGHT	15,150 KG (33,450 LB.)

3

1. NASM's Apollo Lunar Module 2 (LM-2) on exhibit in the Lunar Exploration Vehicles exhibit. 2. Astronaut John W. Young, commander of the *Apollo 16* jumps for the camera and salutes the US flag by the lunar module (LM) *Orion*. 3. With a half-Earth in the background, the lunar module *Eagle* ascent stage with astronauts Neil Armstrong and Edwin "Buzz" Aldrin Jr. approaches for a rendezvous with the Apollo 11 command module manned by Michael Collins.

The lunar module consists of two separate sections. The lower section is the descent stage. Its primary function is to house the fuel and engine necessary for a controlled descent to the Moon's surface. It also contains the landing gear used to keep the vehicle upright, the astronauts' equipment used to explore the Moon and the TV camera used to send back the images of astronauts as they descended the ladder into history.

The upper section of the LM is the ascent stage where the astronauts ate and slept while on the Moon. It contains all of the electronic equipment necessary for controlling the descent stage the ascent stage, and rendezvous. It also contains all the environmental control and life-support systems, the docking apparatus, and the reaction control systems.

It too has an engine, the one used to launch the astronauts from the Moon's surface and into orbit with the command module. The descent stage serves as a launch platform for the ascent stage and is left behind on the Moon.

THE LUNAR ROVING vehicle was the first piloted transportation system designed to operate on the airless, low-gravity terrain of the Moon.

Capable of carrying two astronauts along with their life-support systems, scientific equipment, and lunar samples, the lunar roving vehicle (LRV) greatly extended the area on the Moon that could be explored by humans to about 92 km (57 mi.). This allowed the astronauts to place instruments and collect samples far away from the lunar module.

An LRV was first used by the crew of Apollo 15 on their July 31 to August 2, 1971, mission. The astronauts traversed 25.3 km (15.7 mi.), far exceeding the range of 6.7 km (4.2 mi.) covered by the astronauts on foot during Apollo missions 11, 12, and 14 combined. An LRV was also used during Apollo missions 16 and 17. The three LRVs were driven a total of 88.3 km (54.8 mi.).

Astronauts operated the LRV using the controls at this instrument panel. The foil-covered television camera (visible behind the LRV antenna) was operated by Mission Control back on Earth, which allowed ground controllers to monitor the activities of the astronauts and photograph the launch of the lunar module ascent stage.

The LRV was stored in a compartment on the descent stage of the lunar module (see page 114). The wheels folded under the chassis for storage. Once on the Moon and after

1. The museum's Apollo lunar rover on display in the Apollo to the Moon gallery.
2. Astronaut James B. Irwin readies the lunar roving vehicle (LRV) for exploration of the lunar surface in the Hadley-Apennine region, August 1, 1971. Lunar module *Falcon* is in the background. 3. A detail view of the Apollo lunar rover's control system. 4. Installation of the folded lunar roving vehicle (LRV) in the lunar module at the Kennedy Space Center, April 1971.

DIMENSIONS: —————

LENGTH	3.1 M (10 FT., 2 IN.)
WIDTH	1.8 M (6 FT.)
WEIGHT	210 KG (462 LB.) ON EARTH;
	35 KG (77 LB.) ON THE MOON

releasing storage restraints, astronauts unfolded the rear section and lowered it to the surface. After the front section was similarly prepared, the astronauts released the LRV from the lunar module. The seats were then unfolded and two antennae, the communications unit, and the television camera were attached to the vehicle. After the astronauts loaded their tools on the rear, they were ready to begin operations.

Weighing approximately 209 kg (460 lb.) on Earth, the LRV carried a total payload weight of about 490 kg (1,080 lb.). The LRV had power for up to 78 hours of operation. The vehicle was 310 cm (10 ft., 2 in.) long with a 183 cm (6 ft.) tread width and was 114 cm (44.8 in.) high. The wheelbase was 229 cm (7.5 ft.). Each wheel was individually powered by a ¼-hp electric motor (providing a

total of 1 hp), and the vehicle's top speed was about 13 km/h (8 mph) on a relatively smooth surface.

Two 36-volt batteries provided the vehicle's power, although either battery could power all vehicle systems if required. The front and rear wheels had separate steering systems, but if one steering system failed, it could have been disconnected and the vehicle would have operated with the other system.

The LRV was designed to make several exploration sorties up to a cumulative distance of 65 km (40 mi.). Because of limitations in the astronauts' portable life-support system, however, the vehicle's range was restricted to a radius of about 9.5 km (6 mi.) from the lunar module in case the LRV became immobile and they had to walk back to the lunar module.

ON AUGUST 23, 1977, the *Gossamer Condor* successfully demonstrated sustained and maneuverable human-powered flight and won the £50,000 ($95,000) Kremer Prize.

Originally set at £10,000, the Kremer Prize was established in 1959 by industrialist Henry Kremer. As the years passed with no winners, to create a greater incentive the value was increased until it reached £50,000.

Pilot Bryan Allen took off from Shafter Airport in southern California at 7:30 a.m. and landed 7 minutes, 27.5 seconds later. The official circuit, a figure-8 course around pylons one-half mile apart with a 3-m (10-ft.) hurdle at the beginning and the end, covered 1.85 km (1.15 mi.). Its flight speed was between 16 and 18 km/h (10 and 11 mph) at only ⅓ hp, with Allen, a championship bicyclist and hang-gliding enthusiast, developing the horsepower.

The *Gossamer Condor* was built purely in response to the Kremer Prize. No attempt was made to use the aircraft for aeronautical research or to collect any other data. Designed by Paul B. MacCready and Peter B. S. Lissaman, the *Gossamer Condor* is made of thin aluminum tubes covered with Mylar plastic and braced with stainless steel wires. The leading edges are made of corrugated cardboard and styrene foam. It has a wingspan large enough wing to produce the lift needed to take off and stay airborne at low speed.

The Condor was designed to fly more slowly than earlier human-powered aircraft. This allowed for a lighter, wire-braced machine. The pilot had to pedal longer but expended much less energy and could remain aloft longer and fly farther.

The *Gossamer Condor* was different than other human-powered aircraft in that it could be easily modified or repaired. Safety was not an issue because it flew so low and slow that the pilot was never at risk. After a crash, the *Condor* could be fixed within twenty-

1. The MacCready *Gossamer Condor* hanging on exhibit in Pioneers of Flight gallery of the National Air and Space Museum. 2. The *Gossamer Condor* in flight. 3. The MacCready *Gossamer Condor,* piloted and powered by Bryan Allen, as it approaches marker at Shafter Airport, Shafter, California, August 23, 1977, when this aircraft successfully demonstrated sustained, maneuverable, manpowered flight to win the £50,000 ($95,000) Kremer Prize. 4. A left-side view of the cockpit section.

DIMENSIONS:

WINGSPAN	29.25 M (96 FT.)
LENGTH	9.14 M (30 FT.)
HEIGHT	5.49 M (18 FT.)
WEIGHT	31.75 KG (70 LB.)

four hours, enabling the aircraft to be tested extensively and easily modified. The pilot sat in a semireclining position with both hands free to work the controls. One hand held a handle that controlled both vertical and lateral movement. For turns, the other hand set a lever located beside the seat that controlled wires to twist the wing.

In late 1976 the first flight of the test aircraft was conducted from the parking lot of the Rose Bowl in Pasadena, California. After moving the testing to the Mojave Desert where calmer winds were expected, the first significant flight, one of 40 seconds, took place on December 26. In January 1977 champion cyclist Greg Miller was able to remain aloft for 2.5 minutes. In August the prize-winning flight was made.

In January 1978 the *Gossamer Condor* was donated to the National Air and Space Museum.

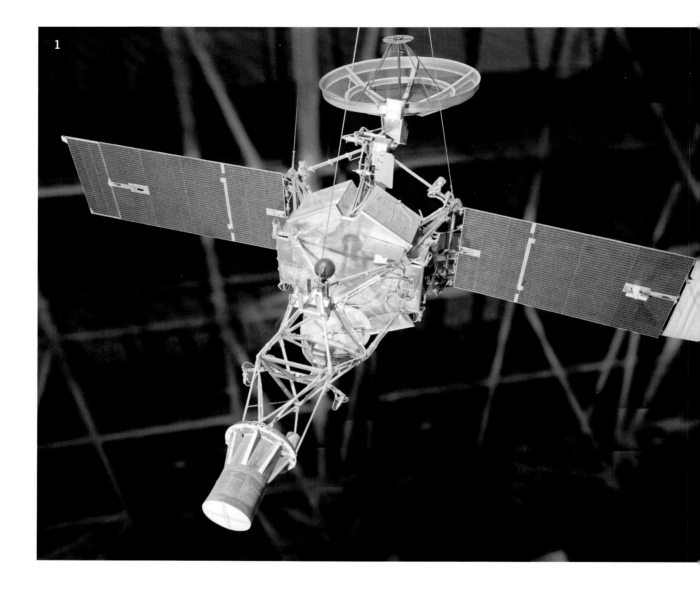

1

THE MARINER 2 full-scale prototype in the National Air and Space Museum is an exact replica of the first spacecraft to radio scientific information about another planet back to Earth.

In the wake of the Soviet Union's Sputnik lunar probes in 1959, the Mariner program was created to send robotic spacecraft to Venus and Mars. Mariner 1 and 2 were designed and built in only eleven months.

Mariner 1 was launched on July 22, 1962, but flew off course and was destroyed after only 293 seconds of flight. Mariner 2 was successfully launched a little over one month later. After a successful mid-course correction was made on September 5, Mariner 2 inexplicably lost attitude control but eventually regained it through the onboard gyroscopes.

The mission continued uneventfully until October 31 when the power from one solar panel suddenly dropped off. Power returned within a week but failed permanently on November 15. Fortunately, Mariner 2 was close enough to the Sun for one solar panel to generate enough electricity to continue the mission.

Approaching Venus from 30 degrees above the dark side, Mariner 2 passed below the planet within 34,773 km (21,600 mi.) on December 14, 1962. Mariner 2 then entered the solar orbit of Venus two weeks later.

DIMENSIONS: _____

HEIGHT	2.5 M (97 IN.)
LENGTH	1.8 M (70 IN.)
WIDTH	1.8 M (70 IN.)
WEIGHT	507 KG (1,118 LB.)

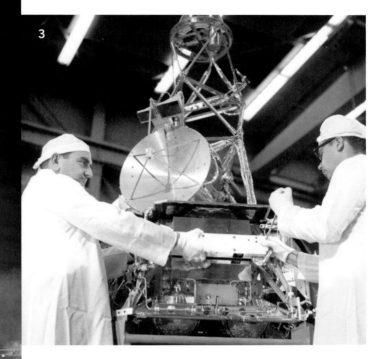

3

1. A view of Mariner 2 hanging on display in the Space Hall, National Air and Space Museum. **2.** A dramatic night launch of an Atlas-Agena 5 carrying the Mariner 1 spacecraft at the Cape Kennedy Launch Complex 12. **3.** Technicians, at work on the Mariner 2 spacecraft at the Jet Propulsion Laboratory (JPL), some time prior to the craft's launch on December 14, 1962.

The instruments onboard Mariner 2 included a magnetometer, charged particle detectors, a cosmic dust collector, and a solar plasma detector. The spacecraft also contained a microwave radiometer and an infrared radiometer to determine the temperature and structure of the Venusian atmosphere.

Mariner 2 was built on a 1.04-m (3.5-ft.) wide hexagonal base that is .36 m (1.2 ft.) thick and houses the spacecraft's electronics and navigational systems. A 3.66-m (12-ft.) tall pyramid-shaped mast contained the experiment packages. The rectangular solar panels that provided power were mounted along the base extending 5.05 m (17 ft.) from the craft. A single large directional dish antenna protruded from an arm away from and below Mariner 2's base.

Contact with Mariner 2 was last made on January 2, 1963. The full-scale prototype on display in the Milestones of Flight gallery was constructed from test components by engineers from NASA's Jet Propulsion Laboratory. It was transferred to the museum in 1975.

THE MARS PATHFINDER conclusively demonstrated the practicability of using innovative ideas and technology to design and build a low-cost spacecraft.

The Mars Pathfinder (MPF) was launched on December 4, 1996, from Cape Canaveral, Florida, by a Delta II booster, and landed on Mars on July 4, 1997. It was the second launch in NASA's Discovery Program and was developed in only three years at a cost of just $150 million. The program was intended to demonstrate the feasibility of building a Mars exploration craft at one-fifteenth the cost of the Viking mission to Mars in 1976. It also demonstrated the use of a rover that independently traversed the surface of the planet.

The MPF consisted of a stationary lander and a separate rover called *Sojourner*. The rover rested on one of the three MPF solar panels that were folded over the spacecraft during launch and interplanetary cruise. After a seven-month flight, the Mars Pathfinder entered the Martian atmosphere traveling at nearly 11,748 km/sec. (7,300 mi./sec.). The entry vehicle's heat shield slowed the spacecraft to 644 km/sec. (400 mi./sec.) in about 160 seconds before deployment of the parachute.

Unique to the Mars Pathfinder was the use of a relatively inexpensive rather than a complicated mechanical landing system. Airbags were deployed approximately eight seconds before impact at an altitude of 300 m (960 ft.). The spacecraft then completed its descent and literally bounced to a safe landing. The MPF landed near the mouth of Ares

1. A view of full-scale model of the Mars Pathfinder lander and a *Sojourner* rover on display in Space Hall at the National Air and Space Museum's Udvar-Hazy Center.
2. A 360-degree panoramic view of the Mars Pathfinder landing site.
3. Engineers test huge, multilobed air bags, which enveloped and protected the Mars Pathfinder spacecraft before it struck the surface of Mars.

DIMENSIONS:

HEIGHT	152.4 CM (60 IN.)
WIDTH	276.9 CM (109 IN.)
DEPTH	320 CM (126 IN.)

Valles, a large outwash plain from one of the largest outflow channel complexes on Mars at 19.33°N, 33.55°W.

After landing, two ramps were unfurled from the solar panel where the rover was mounted. The rover then rolled down one of these ramps and proceeded to take close-up images of the surface using two color cameras mounted on the front and a black-and-white camera on the rear. The rover also contained a rear-mounted alpha proto X-ray spectrometer that provided bulk elemental composition data on surface soils and rocks. In addition, the rover's stubby wheels provided information about the physical characteristics of the surface.

The lander operated for over ninety days, during which time it relayed 2.3 gigabits of data. Data gathered by *Sojourner* was relayed to Earth through the lander. Some of this data suggested the presence of large amounts of water on Mars in the distant past. Communications were lost with the Mars Pathfinder on September 27, 1997.

The Mars Pathfinder in the National Air and Space Museum is a full-scale model that was built by the Jet Propulsion Laboratory. Similar to the actual spacecraft, the rocker-bogie wheels and frame are constructed of aluminum and the scientific, navigation, and communication instruments are made of resin and do not possess all of the detail of the originals.

THE MARTIN B-26B Marauder named *Flak Bait* survived 207 combat missions over Europe, more than any other American aircraft during World War II.

Due to war pressures, the Air Corps went straight into production with the B-26 in 1939, forgoing any prototypes. The consequences were deadly for crews that flew the Martin airplane. The design looked great on paper, but the high wing loading of first production B-26s dramatically increased landing and takeoff speeds. The airplane also had many engine and propeller malfunctions, which led to accidents in training. Production was almost halted on several occasions.

The 22nd Bombardment Group (BG) at Langley Field, Virginia, received the first Marauders in February 1941. Nosewheel strut failures delayed the transition to full operational status, but the first B-26s flew combat missions in the Pacific not long after America entered into World War II. On June 4, 1942, four Army Air Corps Marauders defending Midway Island attacked Japanese aircraft carriers with torpedoes but failed to score hits.

The 319th BG became the first Marauder outfit sent to England. The AAF sent Marauders on to North Africa after the Allies invaded that continent in November 1942. Early Marauders were particularly vulnerable after damage to the hydraulic system. When enemy fire damaged the system, pressure dropped, gravity and airflow forced the bomb bay doors open, and the resulting drag slowed the bomber down and made it easy prey for fighters.

In May 1943 eleven B-26s flew a bombing mission against German forces over Ijmuiden, Holland, to knock out a power station that they had failed to hit three days earlier.

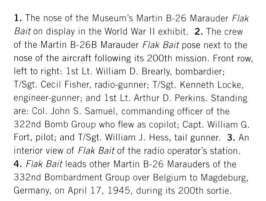

DIMENSIONS:

WINGSPAN	21.63 M (71 FT.)
LENGTH	17.76 M (58 FT., 3 IN.)
HEIGHT	6.55 M (21 FT., 6 IN.)
WEIGHT	10,886 KG (24,000 LB.)

1. The nose of the Museum's Martin B-26 Marauder *Flak Bait* on display in the World War II exhibit. **2.** The crew of the Martin B-26B Marauder *Flak Bait* pose next to the nose of the aircraft following its 200th mission. Front row, left to right: 1st Lt. William D. Brearly, bombardier; T/Sgt. Cecil Fisher, radio-gunner; T/Sgt. Kenneth Locke, engineer-gunner; and 1st Lt. Arthur D. Perkins. Standing are: Col. John S. Samuel, commanding officer of the 322nd Bomb Group who flew as copilot; Capt. William G. Fort, pilot; and T/Sgt. William J. Hess, tail gunner. **3.** An interior view of *Flak Bait* of the radio operator's station. **4.** *Flak Bait* leads other Martin B-26 Marauders of the 332nd Bombardment Group over Belgium to Magdeburg, Germany, on April 17, 1945, during its 200th sortie.

Ten bombers were lost and one aborted. In June all medium bombers were ordered to fly at medium altitudes, about 3,000 m (10,000 ft.). During July 1943, the 3rd Bombardment Wing flew the first Marauder missions without losses, thanks largely to escorting Spitfires (see page 180).

Marauders also operated in the Mediterranean starting in June 1943. The 319th BG pioneered three- and, later, six-ship-abreast formation takeoffs from Decimomannu, Sardinia, in October 1943. During February 1944, the 319th and 320th bombed the Abbey di Monte Cassino. Despite its initial problems, the AAF lost fewer Marauders than any Allied bomber it flew—less than one-half of 1 percent.

Flak Bait, the B-26 in the National Air and Space Museum, was assigned to the 449th Bombardment Squadron, 322nd Bombardment Group and given the identification code "PN-O." Lt. James J. Farrell, who flew more missions in the aircraft than any other pilot, adapted the nickname given his family dog, "Flak Bait," to his aircraft.

This Marauder lived up to its nickname after just a few missions. Other bombers returned unscathed but *Flak Bait* invariably returned full of holes. "It was hit plenty of times, hit all the time," Farrell said. "I guess it was hit more than any other plane in the group."

On April 17, 1945, *Flak Bait* completed its 200th mission, leading the entire 322nd BG to Magdeburg and back. In its career, this bomber logged 725 hours of combat time. It flew two missions on D-Day, 21 missions against V-1 flying bomb launch sites in France, and attacked targets in Holland and Belgium. *Flak Bait* flew three night-bombing missions symbolized by a black bomb symbol painted on the left fuselage below the cockpit.

Few Marauders survive today. The Air Force transferred *Flak Bait* to the Smithsonian in 1949.

SOME AIRCRAFT ARE remembered for the large number produced, others for their length of time in service, and others for their performance. The McDonnell Douglas F-4 Phantom II is remembered for all of these things.

Preliminary design of what was to become the Phantom II began in 1953 as a single-place, long-range attack aircraft designated by McDonnell as the F3H-G. Working closely with the navy, the final outcome was the two-seat F4H-I, the navy's first Mach 2 carrier-based aircraft, capable of carrying missiles, and its primary mission was as an all-weather fleet air defense aircraft, although it retained its original attack capability.

The F4H-I made its first flight on May 27, 1958. One year later the Phantom II joined the fleet. The Phantom II was qualified for both land and sea operations, and within a few years several versions were produced for the US Air Force.

Between 1959 to 1969, the F4H and its derivatives established many altitude and speed records. The earliest version, the F4H-1F, was redesignated the F-4A after 1962. Other navy versions were the F-4B, G, J, N, and S. The US Air Force versions were the F-4C, D, and E. Like the F-4B, the F-4C had no built-in gun but carried Sparrow missiles as its primary attack weapon. An unarmed reconnaissance version was designated the RF-4C. The F-4J was the last fighter version to be placed in quantity production for the US Navy and Marine Corps.

In 1968 the navy chose the F-4J for its Blue Angels flight demonstration team, and in 1969 the air force chose the F-4E for its Thunderbird team. In September 1964 the F-4K was designed for the British Royal Navy, and the F-4M for the Royal Air Force. These versions were equipped with British avionics and Rolls-Royce Spey turbofan engines.

1. NASM's McDonnell F-4S Phantom II on display at the Udvar-Hazy Center. 2. The museum's McDonnell F-4S Phantom II in flight. 3. Vietnam War ace US Navy Lt. Randall "Randy" ("Duke") Cunningham, pilot, and Lt(JG) Willie Driscole, radar intercept officer (RIO), sitting in the cockpit of their F-4S Phantom II. Note the eight "kill" stars. 4. A McDonnell F-4B Phantom II from US Navy Squadron VF-121 on catapult during carrier trials in March 1967.

DIMENSIONS: _____

WINGSPAN	11.6 M (38 FT., 4.75 IN.)
LENGTH	17.7 M (58 FT., 3.75 IN.)
HEIGHT	5 M (17 FT., 4 IN.)
WEIGHT	13,960 KG (30,770 LB.)

Later Iran, South Korea, Spain, Australia, Israel, Japan, Greece, Turkey, and West Germany bought F-4s. The F-4E was the model preferred by overseas air forces. Specially designed for the Japanese was the F-4EJ; it dispensed with most of the offensive systems and was fitted with advanced tail-warning radar and air-to-air guided missiles. The Japanese also ordered the RF-4EJ, an unarmed reconnaissance version. The Germans received the F-4F, which had the air-to-ground weapon delivery system removed to save weight.

Production of the Phantom peaked at a rate of more than 70 aircraft a month, and, by 1979, when production ceased, 5,195 had been built. The last US Navy F-4 made its final carrier landing aboard the USS *America* in October 1986.

The basic F-4B weighs 20,230 kg (44,600 lb.) loaded, has a maximum range of 3,701 km (2,300 mi.), a service ceiling of 18,898 m (62,000 ft.), and a cruising speed of 925 km/h (575 mph), with a maximum speed of 2,390 km/h (1,485 mph) at 14,630 m (48,000 ft.). Two 7,711-kg (17,000-lb) thrust J79-GE-8 turbojets power it.

The McDonnell Douglas F-4S-44 Phantom II in the National Air and Space Museum was assigned to Fighting Squadron 31 (VF-31) stationed at the Naval Air Station, Oceana, Virginia, in 1971. In 1972, VF-31 went aboard the USS *Saratoga,* and entered combat operations on Yankee Station, off the coast of Vietnam. While on a flight on June 21, 1972, the museum's F-4 shot down a MiG-21 with a Sidewinder missile (AIM-9). Its tour in Vietnam included support missions for B-52 raids on Hanoi and Haiphong.

In 1983 the navy converted it from a J model to an S model, which consisted of modernizing the hydraulics, electronics, and wiring, including the installation of leading-edge maneuvering slats. Radar homing and warning (RHAW) antennae and formation tape lights were also added.

When conversion was completed, the F-4S joined Marine Fighter Attack Training Squadron (VMFAT) 101, stationed at Marine Corps Air Station (MCAS), Yuma, Arizona. It was transferred to the Smithsonian in 1988 after it had amassed a total of 5,075 hours of flight time.

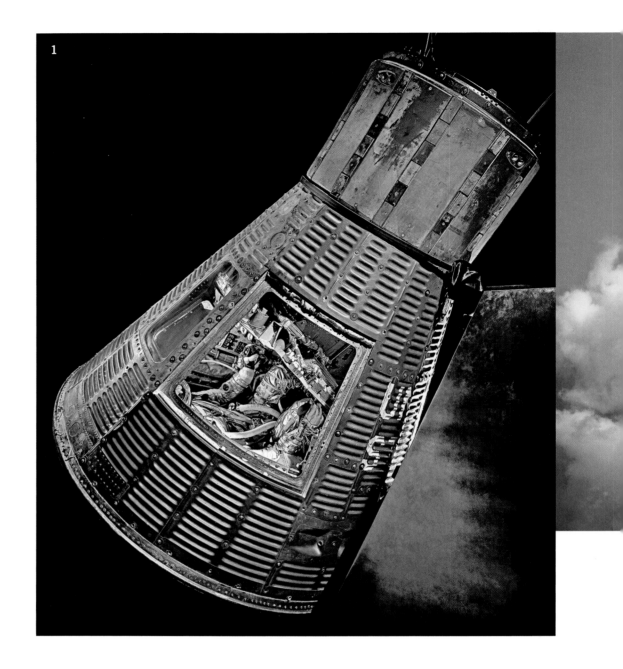

1

ON FEBRUARY 20, 1962, John H. Glenn Jr. became the first American to orbit the Earth in the Mercury spacecraft *Friendship 7*.

The Mercury *Friendship 7* spacecraft was designed to sustain a man in space for several hours, protect him from the heat associated with launch and reentry, allow for control of the spacecraft attitude, provide for observation of Earth and experimentation, and conduct a controlled reentry.

The spacecraft's conical body carried the astronaut, the life-support system, and the electrical power system. The blunt end consisted of an ablative heat shield designed to dissipate the intense heat generated by atmospheric friction during reentry. The cylindrical section at the opposite end of the spacecraft contained the parachutes used during reentry. At the time of launch, the spacecraft weighed 1,935 kg (4,265 lb.), and at recovery it weighed 1,099 kg (2,422 lb.), after the escape tower was jettisoned during launch, and the retrorockets and parachute container were jettisoned during the return to Earth.

DIMENSIONS: ——————

HEIGHT   2.7 M (9 FT.)
MAXIMUM DIAMETER   1.9 M (6 FT., 3 IN.)
WEIGHT   1,300 KG (2,900 LB.)

1. John Glenn's Mercury *Friendship 7* capsule on display in the Milestones of Flight gallery of the National Air and Space Museum. 2. The launch of the Atlas rocket carrying astronaut John H. Glenn and his Mercury MA-6 *Friendship 7* spacecraft at Cape Canaveral, Florida, February 20, 1962. 3. Astronaut John Glenn posed by the *Friendship 7* Mercury capsule.

John H. Glenn Jr., a military pilot with over 9,000 hours of flight time, was chosen to ride in *Friendship 7.* The flight out of the Earth's atmosphere took only 5 minutes. Near the end of his first orbit, Glenn noticed that the spacecraft drifted slowly to the right when the automatic control system was on. He switched to manual control and corrected the problem.

Then an instrument light at mission control indicated that the heat shield and compressed landing bag were loose, which meant that *Friendship 7* and Glenn would be incinerated during reentry if this was not fixed. Fortunately, the retrorocket package was strapped to the heat shield. If the package was retained after retrofire, its straps would hold the heat shield in place. By the time the pack burned away, aerodynamic pressure kept the shield from slipping. All three retrorockets fired while *Friendship 7* was over California, slowing it enough to enter the atmosphere. As

the heat of entry increased, Glenn saw bits of the retropackage fly past his window.

The heat shield did its job. At an altitude of 8,500 m (28,000 ft.), the drogue parachute opened, followed by the main one at 3,000 m (10,000 ft.). Glenn flipped the landing bag release switch and felt a reassuring *clunk* as the compressed landing bag and heat shield dropped into position. The premature deployment signal had been at fault in the ground controller's console.

Friendship 7 splashed down in the Atlantic Ocean 4 hours and 55 minutes after launch. Seventeen minutes later, the destroyer USS *Noa* was floating alongside, ready to retrieve the bobbing spacecraft. Once the craft was cradled in a mattress pallet on *Noa*'s deck, Glenn fired the explosive bolts holding the hatch in place. A hot, tired astronaut emerged from the spacecraft an America hero.

NASA transferred Mercury 7 to the museum in 1962.

1

THE STORY OF air combat over Europe cannot be told without great emphasis on the Messerschmitt Bf 109.

The Bf 109's first public demonstration took place at the 1936 Olympic Games held in Berlin, but the plane's first real impact on the aviation world came during the international flying meet held in Zurich in the summer of 1937. There, five Bf 109s demonstrated the plane's outstanding speed, climbing, diving, and maneuverability.

Twenty-four Messerschmitt fighters were later delivered to Spain for the German-manned Condor Legion during the Spanish Civil War. By the time England declared war on Germany, the Messerschmitt was being mass-produced in the Bf 109E series and was ready to enter the fight.

The Spitfire, the Bf 109's first major opponent, was slightly faster and definitely more maneuverable, but its performance at altitude was inferior. There was also little difference in pilot skill between the Luftwaffe and the Royal Air Force (RAF), although pilots in the RAF had the advantage of fighting over their own country, while the critical range of the Bf 109s limited German fighting time to about twenty minutes.

As Allied bomber formations and fighter-bombers pushed the war into Germany, the Bf 109s were forced into a combat role for which they were not designed—close ground support. In this capacity the Bf 109s were heavily battered by Allied fighters and ground fire. The Messerschmitt also relentlessly attacked the massive bomber formations, only to be heavily pounded by the bombers' defensive crossfire. In every air encounter over Europe, the Bf 109s could generally be counted on to appear for a fight.

DIMENSIONS:

WINGSPAN	9.92 M (32 FT., 6.5 IN.)
LENGTH	9.02 M (29 FT., 7 IN.)
HEIGHT	3.4 M (11 FT., 2 IN.)
WEIGHT	2,700 KG (5,953 LB.)

3

4

1. A view of the museum's Messerschmitt Bf 109G-6 following the completion of its restoration at the Paul E. Garber Facility in 1974. **2.** A Messerschmitt Bf 109G-6 of 7 Squadron III./JG 27 flying over the eastern Mediterranean in late 1943. **3.** A Messerschmitt Bf 109G (Gustav) on the ground at an unidentified Luftwaffe airfield. **4.** Possibly the museum's Messerschmitt Bf 109G undergoing static testing at Building 23, Wright Field, Ohio, during World War II.

As new and improved models of Allied fighters entered the war, the Germans countered with upgraded models of the Bf 109, primarily increasing power rating in the Daimler-Benz (DB) engine. When German production stopped, the G series of the Bf 109 was produced in far greater numbers than any other model, with 21,000 being completed by the end of 1944. Known as "Gustav," the Bf 109G was powered by a DB 605 engine. The airplane had two MG 131 machine guns, a single 30-mm MK 108 cannon firing through the spinner, and sometimes carried two underwing MG 151/20 weapons. This combination was ideal for bomber interception but severely reduced the Bf 109's efficiency in fighter-versus-fighter combat.

The Bf 109 in the National Air and Space Museum was shipped to the United States for evaluation with a number of other German aircraft near the end of the war. It was stripped of all its unit markings, camouflage, and even its serial number. The markings and two-tone gray camouflage pattern currently on the airplane are for number 2 of the 7th Squadron, 3d Group, 27th Wing that operated in the eastern Mediterranean in late 1943 as an escort fighter. It was transferred from the US Air Force to NASM in 1948.

ALTHOUGH NOT A significant factor in World War II, the Me 262 was the world's first operational jet fighter.

The Me 262 Schwalbe was powered by two Junkers Jumo 004 B turbine engines and had sleekly swept wings and four 30-mm cannons. With a top speed of approximately 869.4 km/h (540 mph), the Me 262 was 193.2 km/h (120 mph) faster than the famed North American Mustang at the same altitude.

The Me 262 introduced many features found on later aircraft, including the swept wing, wing slots, underslung engine nacelle, and heavy cannon armament mounted in the nose. Me 262s were reportedly delightful to fly, as long as the pilot used care in moving the throttles to avoid an engine compressor stall.

Conceived in 1938, the Me 262 was designed by a team led by Dr. Waldemar Voigt. It went through a long gestation period, not making its first flight until April 18, 1941, and then only under the power of a Junkers Jumo 210G piston engine of about 700 horsepower. Jet engine development, although more advanced in Germany than elsewhere, was still in a primitive state, and the turbine engines intended for the sleek fighter were not ready. On July 19, 1942, Flugkapitän Fritz Wendel made the first takeoff under jet power, and from that point on the Me 262 became a ray of hope in the increasingly dark skies of the German Luftwaffe.

2

DIMENSIONS: ════

WINGSPAN	12.48 M (40 FT., 11.5 IN.)
LENGTH	12.13 M (39 FT., 9.5 IN.)
HEIGHT	3.84 M (12 FT., 7 IN.)
WEIGHT	4,419 KG (9,742 LB.)

3

4

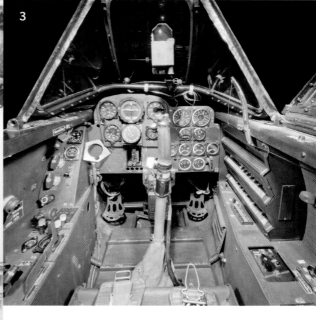

1. The museum's Messerschmitt Me 262A-1a following the completion of its restoration at the Paul E. Garber Facility in March 1979. 2. A view of the prototype Me 262V3. This was the first all-jet Me 262 to fly. 3. A view of the cockpit of NASM's Messerschmitt Me 262A-1a Schwalbe. 4. A US-captured Messerschmitt Me 262 FE 110 flown by Col. Harold Watson makes a low pass.

For a time historians believed that Adolf Hitler's order to build the Me 262 solely as a bomber delayed its introduction into combat as a fighter interceptor. This is not the case. Rather, jet engine development proved lengthy and difficult. Hitler's order did divert some 30 percent of production airframes to the Me 262A-2a Sturmvogel (Stormbird) bomber type.

The vastly superior performance of the Me 262 gave confidence to the fortunate pilots who flew them, but the Allied dominance of the air was so complete that the Schwalbe never reached its full potential. The airfields from which it flew were under constant attack, and, in the last days of the war, the remaining Me 262s were forced to operate from makeshift bases constructed along Germany's famous autobahns. Although 1,443 Me 262s were completed, it is estimated that only about 300 saw combat.

The Me 262 in the National Air and Space Museum was captured at Lechfeld, Germany, by the US Air Force and brought to the United States for testing. It was donated to the Smithsonian in 1950, and after 6,077 man-hours of restoration, the aircraft appeared with the unit insignia and victory markings of the JG 7 (Fighter Wing 7).

AN ANSWER TO Soviet premier Joseph Stalin's urgent call for a high-altitude day interceptor, the MiG-15 shocked the West with its capabilities and made the acronym MiG synonymous with "Soviet fighter plane."

The MiG-15 was the first Soviet jet to benefit from the new Rolls-Royce Nene and Derwent jet engines, which were sold by Britain to the USSR. These were immediately copied and refined by the Soviets, and as the RD-500, Klimov RD-45, and modified VK-1, they gave a powerful boost to Soviet jet technology.

Designed in 1946, the MiG-15 was the first Russian airplane to feature a swept wing, a pressurized cockpit, and an ejection seat. At a 35-degree sweep, the MiG-15's mid-wing configuration minimized drag. To minimize the outward airflow along the swept wing, two wing fences were fitted to the upper surface of each wing. A one-piece Fowler flap was installed in each wing, and air brakes were fitted to the rear fuselage under the swept cruciform tail.

A compact fuselage of circular cross section provided for minimum frontal area and maximum space utilization. Fuel was contained only in the fuselage, and the main retractable landing gear was built into the wings. The entire airframe was stressed to withstand an 8g load. The heavy armament of a single 37-mm cannon and two 23-mm cannons was designed specifically to destroy large bombers, but they proved to be devastatingly effective against smaller fighters. The guns were installed in an ingenious retractable tray under the aircraft's nose that could be raised and lowered into the aircraft in less than three minutes allowing for rapid rearmament and maintenance.

1. The museum's MiG-15 on display at the Udvar-Hazy Center.
2. A Mikoyan-Gurevich MiG-15 (NATO code name Fagot) in flight. Notice that markings on right side of nose have been blocked out.
3. Chinese ground crew prepares two MiG-15s for a mission.
4. A close-up view of the nose of the museum's MiG-15.

DIMENSIONS:

WINGSPAN	10.08 M (33 FT., 1.5 IN.)
LENGTH	10.1 M (33 FT., 1.75 IN.)
HEIGHT	3.37 M (10 FT., 10 IN.)
WEIGHT	3,523 KG (7,767 LB.)

MiG-15 production was authorized in March 1948, only 3 months after the first test flight, and substantial numbers were in service by the end of 1948 with both Soviet air forces, the VVS (the tactical air arm) and IA-PVO (the air defense arm).

Late in 1950, MiG-15s piloted by Russians appeared over North Korea, and their deadly attacks, using one 37-mm and two 23-mm cannons, quickly ran all piston-engined aircraft from the skies during daylight hours, including the B-29. First-generation jets like the F-80 and F-84 were no match, and the United States had to rush the North American F-86 Sabre (see page 142) into Korea to reestablish air superiority.

Despite its high speed, excellent maneuverability, and high service ceiling, the MiG-15 was not very stable as a gun platform, and it had a tendency to Dutch roll at high speeds because of wing flexing and poor aileron effectiveness. Its cockpit instrumentation was primitive and stick forces were heavy. In combat against the F-86, a

much more advanced fighter but with very similar performance, the MiG-15 suffered an estimated 10-to-1 loss ratio, although much of this disparity must be attributed to the superior training of the US pilots, especially when compared with the inexperienced Chinese and North Korean pilots.

The Chinese and North Koreans had MiG-15s before the Korean War ended, and the MiG-15 was ultimately flown in some thirty-five countries, remaining in service in China as late as 1978, where it was called the J-2 (F-2 in an export version). The MiG-15UTI trainer version, also used throughout the world, is still in service today.

More than 12,000 MiG-15s were built in seventeen versions, in Poland, Czechoslovakia, and China, as well as in the USSR. Many Chinese F-2s have made their way to the United States, where they can be seen at air shows.

The MiG-15 on display at the National Air and Space Museum is a Chinese F-2 acquired in September 1985.

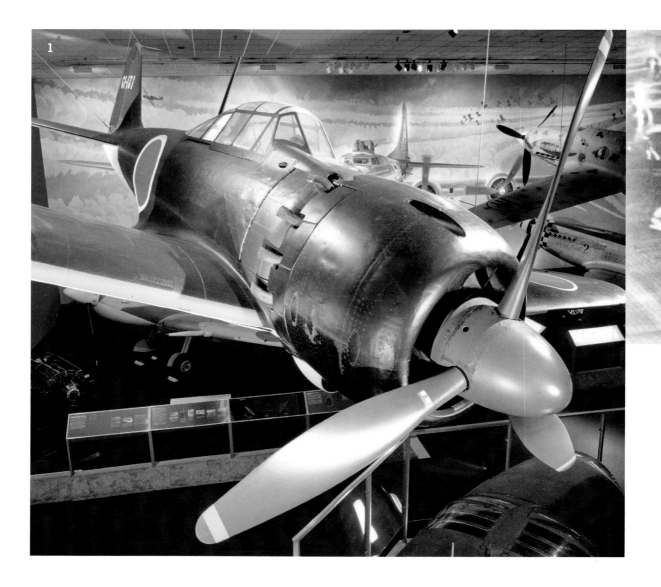

THE ZERO (MITSUBISHI A6M5 Reisen) was the symbol of Japanese airpower during World War II.

The Zero first flew combat missions in China in July 1940, and Zero pilots felled nearly 100 Chinese aircraft and lost only 2 Zeros. The Japanese flew 328 Zeros against American forces at Pearl Harbor and the fall of the Philippines Code-named "Zeke" by the Americans, the Zeros totally outclassed all Allied fighter aircraft for the first six months of the war.

Key to the Zero's performance was weight. A new lightweight aluminum alloy developed in Japan was used for the airframe, and there was no armor plate or self-sealing fuel tanks, eliminating hundreds of kilograms.

The Zero could climb faster and outmaneuver Allied fighters in dogfights. But when American pilots changed their tactics and avoided close combat, the Zero lost its advantage. Flying Grumman F4F Wildcats, the Americans also attacked in single, straight passes with guns blazing, which turned out to be quite effective against the Zero.

Ultimately, the United States built the Lockheed P-38 Lightning and the Grumman F6F-3 Hellcat (see page 82). Both of these airplanes had enough speed and climb to engage or avoid the Zero.

WINGSPAN	11 M (36 FT., 1 IN.)
LENGTH	9.12 M (29 FT., 11 IN.)
HEIGHT	3.51 M (11 FT., 6 IN.)
WEIGHT	1,876 KG (4,136 LB.)

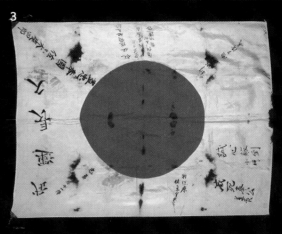

1. A close-up view of the museum's Mitsubishi A6M5 Reisen (Zero Fighter) "Zeke" hanging on display in the World War II Aviation exhibit. **2.** Mitsubishi A6M2 Model 21s warm up on the deck of a Japanese aircraft carrier in preparation for the attack on Pearl Harbor, December 7, 1941. **3.** A captured Japanese flag. The families of Japanese servicemen would write messages on flags, which the servicemen would take into battle for luck and protection. **4.** A US-captured Mitsubishi A6M5 Reisen ("Zeke") in flight; number 29 on tail and cowling; shows both US and Japanese markings.

The Zero remained in production throughout the war. The Japanese built a total of 11,291 Zeros in many versions, more than any other warplane.

The Mitsubishi A6M5 Model 52 on display in the National Air and Space Museum probably came from a group of Japanese aircraft captured on Saipan Island in 1944. Interestingly, during the restoration process, technicians discovered several Japanese messages inscribed inside the metal skin of the aft fuselage, translated as: "Pray for absolute victory," "Win the air war," "Devastation of the American and the British," and "New Year's Day, 1944." The outside markings are from the 261st Naval Air Corps, commonly known as the Tiger Corps, which participated in some of the fiercest fighting of the war. It was transferred from the US Air Force in 1946.

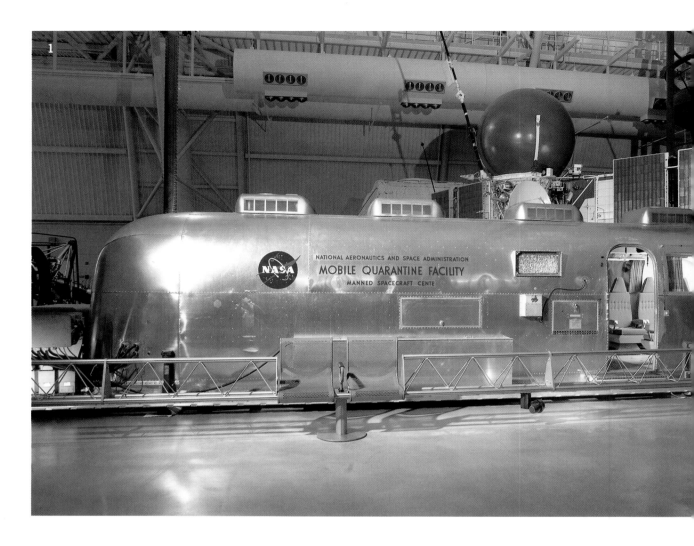

THE MOBILE QUARANTINE Facility isolated astronauts from contact with other people in order to prevent the spread of any lunar-based contagions.

The facility was created by NASA contract engineers, who converted a standard Airstream recreational trailer (consisting of living and sleeping quarters, a kitchen, and a latrine) by sealing it and installing equipment to maintain negative internal pressure through the filtration of effluent air. Other modifications included internal backup power, self-contained air-conditioning and communications equipment, and medical diagnostic equipment. There was enough food and supplies to accommodate six people for up to ten days.

After splashdown in the Pacific Ocean, the astronauts donned biological isolation garments that were designed to prevent the spread of any alien organisms they might have acquired during their journey or while on the Moon. The astronauts breathed through a respirator and mask. They were then flown by helicopter to an awaiting aircraft carrier and transferred to the Mobile Quarantine Facility (MFQ).

Once in the MQF, the astronauts remained there for 65 hours, and were transported, while still in the MQF, by ship, aircraft, and truck to the Lunar Receiving Laboratory (LRL) at the Johnson Space Center in Houston. In Houston they walked through a plastic tunnel that connected the MQF to the LRL. The astronauts were then required to stay in medical isolation for 21 days.

HEIGHT	2.61 M (8.6 FT.)
WIDTH	2.74 (8.9 FT.)
LENGTH	10.66 M (35 FT.)
WEIGHT	5,669.9 KG (12,500 LB.)

1. Apollo Mobile Quarantine Facility (MQF) on display at Udvar-Hazy Center. **2.** Apollo 11 flight crew members Neil A. Armstrong, Michael Collins, and Edwin E. "Buzz" Aldrin Jr. inside the MQF, being greeted through the window by President Richard M. Nixon, July 24, 1969. **3.** Edwin E. "Buzz" Aldrin Jr., Neil A. Armstrong, and Michael Collins wave from the USS *Hornet* as they walk from their recovery helicopter to the Mobile Quarantine Facility, after completing the Apollo 11 mission, July 24, 1969. **4.** Astronauts Neil Armstrong, Edwin "Buzz" Aldrin Jr., and Michael Collins in the Mobile Quarantine Facility after the Apollo 11 flight, 1969.

Fortunately, exhaustive tests on the astronauts and the lunar samples uncovered no life-forms, and this postflight procedure was dropped after the Apollo 14 mission.

The Mobile Quarantine Facility in the National Air and Space Museum is one of four built for NASA, and was used by the crew of *Apollo 11* after their return to Earth. NASA transferred the MQF to the Smithsonian in 1974.

THE NIEUPORT 28C1 was developed in mid-1917 and was the first biplane fighter design produced

by Nieuport that had relatively equal-chord upper and lower wings, a departure from the
highly successful earlier line of sesquiplane "V" strut single-seat scouts, the most famous of
which were the Nieuport 11 and the Nieuport 17.

In an attempt to compete with the superior performance of the Spad VII and the
recently introduced Spad XIII (see page 178), Nieuport explored the use of the more
powerful and heavier 160-hp Gnôme rotary engine. The surface area of the lower wing of
the Nieuport 28 was increased to compensate for the greater weight of the new engine.

The French Air Service rejected the new Nieuport design as a frontline fighter in favor
of the sturdier, more advanced Spad XIII. However, the Nieuport 28 found a place with the
new American squadrons. Having no suitable fighter design of its own, the United States
adopted the Nieuport 28 as a stopgap measure before the much-in-demand Spad XIIIs
could be made available from the French. The Nieuport 28 performed creditably as the first
operational pursuit aircraft in the fledgling US Air Service of the American Expeditionary
Force. Thus, the primary significance of the Nieuport 28 is that it was the first fighter aircraft
to serve with an American fighter unit under American command and in support of US
troops.

The Nieuport 28 was also first type to score an aerial victory with an American unit.
On April 14, 1918, Lts. Alan Winslow and Douglas Campbell of the "Hat-in-the-Ring" 94th
Aero Squadron, both piloting Nieuport 28s, each downed an enemy aircraft in a fight that
took place directly over their home airfield at Gengoult.

The Nieuport 28 made its mark in aviation history after World War I as well. Of the

DIMENSIONS:

WINGSPAN	8.2 M (26 FT., 11 IN.)
LENGTH	6.5 M (21 FT., 4 IN.)
HEIGHT	2.5 M (8 FT., 2 IN.)
WEIGHT	533 KG (1,173 LB.)

1. The museum's Nieuport 28C1 on display at the Udvar-Hazy Center. **2.** James Meissner and his Nieuport 28 fighter no. "8" of the 94th Aero Squadron on May 15, 1918, following a decoration ceremony. Meissner's personal markings are replicated on the museum's aircraft. **3.** A United States' Navy Nieuport 28 on a launching platform built out over the forward turret of a battleship for shipboard launching trials, circa 1919 to 1921.

297 total Nieuport 28 fighters procured by the United States from the French government during World War I, 88 were returned to the United States after the war. Twelve Nieuports, along with examples of several other European types brought back, were used by the US Navy from 1919 to 1921 for shipboard launching trials. Many, often harrowing, launches were undertaken. Some of the 12 Navy Nieuport 28s were destroyed in accidents. The surviving aircraft, worn out beyond repair, were surplused after the trials. The other 76 Nieuport 28s that were brought back to the United States after the war were operated by the US Army in the 1920s at various bases and airfields such as McCook, Mitchel, and Bolling Fields.

Several of the surviving Nieuports found their way into various private hands. Some were modified for air racing, but a number found their way into Hollywood movies, most notably in the famous *Dawn Patrol* films of 1930 and 1938. Still others became privately owned airplanes flying in various sporting and commercial capacities.

The Nieuport 28 in the National Air and Space Museum is an assemblage of components of various aircraft that were all manufactured at the end of or soon after World War I. In light of the vague provenance of the museum's Nieuport, some considered judgment was required concerning the final configuration and markings of the aircraft when it was restored by the museum.

Even though the museum's Nieuport is not a war veteran, because it was manufactured after the United States ceased to use the aircraft in combat, the best alternative was to configure the airplane in this fashion. The particular Nieuport 28C1 that the museum chose to represent was that of 1st Lt. James A. Meissner, a 5⅔ victory ace of the 94th Aero Squadron. This aircraft was chosen, rather than one of the more famous ones, such as Eddie Rickenbacker's, Douglas Campbell's, or Alan Winslow's, because it is representative of the famous "Hat-in-the-Ring" 94th Aero Squadron without misleading museum visitors into thinking that the NASM aircraft is actually one of the especially well-known American Nieuport 28s.

THE F-86 SABRE joined the ranks of the great fighter aircraft during combat operations high above the Yalu River during the Korean Conflict from 1950 to 1953.

The F-86 Sabre is built around a wing that is swept back at an angle of 35 degrees. On December 28, 1947, the US Air Force ordered 221 P-86As, which incorporated this wing design. It was powered by the 4,850-lb. thrust General Electric J47-GE-1 engine. In June 1948, a month after the first P-86A flight, its designation was changed to F-86A.

On September 15, 1948, an F-86A set a world speed record of 1,080 km/h (671 mph). In addition to its high performance, the F-86A had excellent handling characteristics and was well liked by its pilots. The Sabre was armed with six .50-caliber M3 machine guns mounted in the nose. Despite the effectiveness of these weapons during World War II, they were not as effective as cannons against opposing jet fighters in the skies above Korea. However, the Mark 18 manual-ranging computing gun sight was replaced in later models with the highly effective A-1CM, which used radar ranging and thereby greatly improved accuracy and lethality.

In December 1950 the 4th Fighter Interceptor Wing, one of the first of the US Air Force's Sabre units, arrived in Seoul to fight the Soviet-built swept-wing Mikoyan-Gurevich MiG-15 (see page 134). On December 17, in the first known combat between swept-wing fighters, Lt. Col. Bruce H. Hinton shot down a MiG-15. By the end of the Korean War, the Sabres had destroyed almost 800 MiG-15s with the loss of fewer than 80 F-86s.

4 MIG ALLEY 200 MILES

1. The museum's North American F-86A Sabre on display at the Udvar-Hazy Center. **2.** A North American F-86 Sabre launches 5-inch rockets at a target range at Nellis Air Force Base, Nevada. **3.** North American F-86 Sabres undergoing maintenance, at an air force facility in Korea, circa 1952–54. **4.** An F-86 pilot returns from a mission to "MiG Alley."

DIMENSIONS:

WINGSPAN	11.3 M (37 FT., 1 IN.)
LENGTH	11.5 M (37 FT., 6 IN.)
HEIGHT	4.5 M (14 FT., 8 IN.)
WEIGHT	4,750 KG (10,495 LB.)

Although the MiG-15 was faster, possessed a higher climb rate and service ceiling, and carried a more powerful armament, the F-86 proved to be a superior gun platform. This was largely the result of the National Advisory Committee for Aeronautics and the US Air Force's secret X-1 program, which developed a full flying tail that smoothed the transonic airflow, allowing the Sabre to fly much more steadily and effectively.

The F-86 progressed through several improved versions—the F-86E, F, H, D, and K models. The changes, in most cases, included improved armament, more powerful engines, and control-system modifications. The F-86D, however, was an all-weather interceptor with a radar nose, and was armed with rockets instead of machine guns. The F-86K was a D-model with 20-mm machine guns instead of rockets. Some F-86s were built under license in Canada, Japan, and Italy. Of the 8,443 Sabres produced, 554 were F-86As.

The National Air and Space Museum's F-86A Sabre was assigned to the 4th Fighter Interceptor Group and flew combat missions against MiG-15s near Seoul. It is displayed in the markings of the 4th Fighter Wing, the first F-86 unit in Korea. It was transferred from the US Air Force in 1962.

"NEARLY EVERY FLIGHT that was made by *Excalibur III* broke some kind of record," said this Mustang's last pilot and owner, Capt. Charles F. Blair Jr.

After World War II, the P-51C Mustang, now in the National Air and Space Museum, was sold as surplus to A. Paul Mantz. A movie stunt and race pilot, Mantz planned to enter the cross-country Bendix Air Race from the West Coast to the site of the National Air Races in Cleveland, Ohio. To eliminate the need for an intermediate stop, he converted the wing into a large fuel tank by sealing the interior, which more than doubled the aircraft's range.

In 1946 and 1947, Mantz's P-51 came in first in the Bendix Air Races. In 1948 it came in second and in 1949 it finished third, flown by hired pilots Linton Carney and Herman "Fish" Salmon, respectively. In 1947, Mantz set a coast-to-coast speed record in each direction with the Mustang, then called *Blaze of Noon*.

Following its last Bendix Race, Charles F. Blair bought the Mustang to set a solo, round-the-world speed record. Blair was an experienced captain with Pan American World Airways at the time, and had established his reputation by setting records in flying boats during his numerous crossings of the Atlantic during World War II.

After the Korean War began, Blair had to change his plans because flying across international borders in a combat plane during wartime would not have been prudent. New plans were set for the plane that Blair had renamed *Excalibur III,* from the Sikorsky VS-44 Excalibur Flying Boat that he flew for American Export Airlines. After careful preparations, Blair flew his Mustang from New York to London on January 31, 1951, in 7 hours and 48 minutes, breaking the existing speed record by 1 hour and 7 minutes. This record stands today for reciprocating-engine, propeller-driven airplanes.

DIMENSIONS:

WINGSPAN	11.28 M (37 FT., 3 IN.)
LENGTH	9.83 M (32 FT., 3 IN.)
HEIGHT	3.89 M (12 FT., 9.5 IN.)
WEIGHT	4,445 KG (9,800 LB.)

1. The North American P-51C *Excalibur III* following its restoration at the Paul E. Garber Facility in January 1978. **2.** The pilot Charles Francis Blair Jr. in the cockpit of *Excalibur III*. **3.** A view of a North American P-51D Mustang *Tika IV* in flight with Lt. Vernon L. Richards at the controls; somewhere over England, circa 1944–45.

Blair had developed a new method of air navigation in polar regions, where the magnetic compass is unreliable, if not useless. Blair believed that navigation was possible by plotting sun lines at predetermined locations and times. To prove his theory, he left Bardufoss, Norway, in *Excalibur III* on May 29, 1951, heading north to Fairbanks, Alaska, via the North Pole. There were no intermediate emergency landing points and no communications or radio navigation aids available to him after departing Norway. Exactly as planned, 10 hours and 27 minutes after takeoff, *Excalibur III* arrived at Fairbanks. For this accomplishment, Blair was awarded the Harmon International Trophy in 1952 by President Harry Truman.

On the return trip, *Excalibur III* also carried the first official air mail across the North Pole from Fairbanks to New York, and a record was set for the first nonstop transcontinental solo crossing of the Alaska-Canadian route from Fairbanks to New York City.

During World War II, P-51 squadrons accompanied 8th Air Force bombers on long-range raids over Europe.

Bomber losses dropped sharply when the Mustangs protected them. The Mustang also proved itself an excellent Boeing B-29 Superfortress (see page 28) escort in the Pacific Theater.

The US Air Force used the airplane during the Korean War for close-support missions, but it was not suited for this dangerous mission. One of the Mustang's few vulnerable spots was its cooling system. A single bullet through a radiator or pipe was usually enough to down a P-51.

North American built more than 14,000 Mustangs and more D-models (8,302) than all other variants combined. The most significant D-model features were a rear fuselage reduced in height to accommodate a new bubble canopy and an increase in armament from four to six .50-caliber machine guns. The US Air Force did not withdraw P-51s from service until 1957. The museum also has a D model in the collection.

Excalibur III was donated to the Smithsonian in 1953. It was completely restored in 1977.

1

THE NORTH AMERICAN X-15 rocket-powered research aircraft bridged the gap between manned flight within the atmosphere and manned flight beyond the atmosphere into space.

Development of the X-15 began in 1954 in a joint research program sponsored by the National Advisory Committee for Aeronautics (forerunner of NASA), the US Air Force, the US Navy, and private industry. North American was selected as the prime contractor that would build the aircraft.

The X-15 became the first winged aircraft to attain velocities of Mach 4, 5, and 6. Because of its high-speed capability, the X-15 was designed to withstand aerodynamic temperatures on the order of 649°C (1,200°F). The aircraft was fabricated with a special high-strength nickel alloy named Inconel X.

Air-launched from a modified Boeing B-52 Stratofortress, the X-15 required conventional aerodynamic control surfaces to operate within the atmosphere and special thruster reaction control rockets located in the nose and wings that enabled the pilot to maintain control when flying on the fringes of space. Because of the potential dangers to the pilot should the X-15's cockpit lose pressure, X-15 pilots wore specially developed full-pressure protection spacesuits.

The Air Force and the National Advisory Committee for Aeronautics developed a special 485-mile-long test corridor stretching from Wendover Air Force Base, Utah, to Edwards Air Force Base, California. It was planned that the X-15 would be air-launched from a Boeing B-52 near Wendover, then fly down this corridor—the High Range— to Edwards.

1. The museum's North American X-15 hanging on public display in the Milestones of Flight gallery.
2. A North American X-15 mounted on pylon beneath the right wing of its Boeing NB-52 Stratofortress mother ship shortly before being dropped by the Stratofortress.
3. The North American X-15 in flight moments after being dropped from the Boeing NB-52 Stratofortress mother ship.   4. An X-15 on the ground, being secured by crew members after a research flight; in the background the Boeing B-52 mother ship used for launching makes a low-altitude fly-by.

DIMENSIONS: ─────

WINGSPAN	6.82 M (22 FT., 4 IN.)
LENGTH	15.47 M (50 FT., 7 IN.)
HEIGHT	3.96 M (13 FT.)
WEIGHT	5,670 KG (12,500 LB.)

Each flight was monitored by tracking stations at Ely and Beatty, Nevada, and at Edwards. The range lay along a series of flat dry lakes where the X-15 could make an emergency landing if necessary. Nothing this extensive had previously existed in flight research, and it foreshadowed the worldwide tracking network developed by American-manned spacecraft ventures. The X-15 would complete its research mission and then, followed by special Lockheed F-104 chase aircraft, land on the hard clay of Rogers (formerly Muroc) Dry Lake. Because the X-15 featured a cruciform tail surface arrangement, it was necessary for the designers to make the lower half of the ventral fin jettisonable prior to landing so that the conventional two-wheel, nose-landing gear and two tail-mounted landing skids could support the aircraft.

Three X-15 research aircraft were built, and they completed 199 research flights. The National Air and Space Museum has X-15 #1. The fastest X-15, #2, was rebuilt following a landing accident as the advanced X-15A-2 with increased propellant capacity. X-15 #3 featured an advanced cockpit display panel and a special adaptive control system. The aircraft made many noteworthy flights until it crashed during atmospheric reentry, following pilot disorientation and a control-system failure. The pilot, Capt. Michael Adams, was killed.

The X-15 flew to a peak altitude of 108 km (67 mi.), and the X-15A-2 attained a speed of Mach 6.72, or 7,297 km/h (4,534 mph).

The X-15 spearheaded research in a variety of areas, including hypersonic aerodynamics, winged reentry from space, life-support systems for spacecraft, aerodynamic heating and heat transfer research, and earth sciences experiments. The X-15A-1 was transferred to the museum from NASA in 1969.

ON NOVEMBER 23, 1935, explorer Lincoln Ellsworth, with Canadian pilot Herbert Hollick-Kenyon, took off in the Northrop Gamma *Polar Star* from Dundee Island in the Weddell Sea and headed across Antarctica to Little America.

Antarctica was the last continent to be discovered and the only one that was mapped entirely from the air. Aerial explorers from the United States, Great Britain, Australia, Norway, Canada, and France can be credited with this feat, and Ellsworth was one of the most tenacious of these explorers.

Ellsworth first took the *Polar Star* to the Antarctic in 1934. Sir Hubert Wilkins, the famous Australian polar explorer, went along as advisor, and Bernt Balchen was the pilot. However, the 4.6-m (15-ft.) thick ice on which the *Polar Star* was standing broke apart, and one of the skis slipped through a crack. The aircraft was almost lost, but after hours of work it was recovered and put back on the ship to be returned to the United States for repairs.

In November 1935, Ellsworth and Hollick-Kenyon finally succeeded in flying the *Polar Star* across Antarctica. After their takeoff on the 23rd, they flew at an altitude of 4,084 m (13,400 ft.) to cross the 3,658-m (12,000-ft.) peaks of the Eternity Range. Ellsworth named a portion of that area James W. Ellsworth Land in honor of his father.

The *Polar Star* made four landings during its flight across the Antarctic. After a blizzard that occurred during the night at the third camp, the inside of the plane was packed

DIMENSIONS:

WINGSPAN	14.6 M (48 FT.)
LENGTH	9 M (29 FT., 9 IN.)
HEIGHT	2.7 M (9 FT.)
WEIGHT	1,589 KG (3,500 LB.)

1. The Northrop Gamma 2B *Polar Star* on exhibit in the museum's Golden Age of Flight gallery. **2.** The Northrop Gamma 2B *Polar Star* on snow-covered ice, presumably taken during Lincoln Ellsworth's 1935 trans-Antarctic flight. **3.** Jack Northrop, Bernt Balchen, Lincoln Ellsworth, and an unidentified man pose standing beside Northrop Gamma 2B *Polar Star* at the Northrop Aircraft factory, October 1932. **4.** The Northrop Gamma 2B *Polar Star* on floats.

solid with drifted snow. The two explorers spent a whole day scooping out the dry, powdery snow with a teacup.

On December 5, fuel exhaustion forced them down about 40 km (25 mi.) short of their goal of Little America. They walked for six days and then settled down in the camp abandoned by Richard E. Byrd several years earlier. The British Research Society ship *Discovery II* sighted them on January 15, 1936. Hollick-Kenyon later returned to recover the *Polar Star*.

The total distance flown by the *Polar Star* before its forced landing was about 3,862 km (2,400 mi.). The US Congress voted Ellsworth a special gold medal for his Antarctic exploration and "for claiming on behalf of the United States approximately 350,000 square miles of land

in the Antarctic representing the last unclaimed territory in the world."

The *Polar Star* was one of two Gammas produced in 1933 by the newly established Northrop Corporation. The Gamma is a low-wing, all-metal cantilever monoplane with a 710-hp, 9-cylinder Pratt & Whitney Hornet engine. The one built for Ellsworth had two seats in tandem with dual controls. The other Gamma was built for Frank Hawks, at the time a pilot for Texaco. His Gamma was a single-seat model. On June 2, 1933, Hawks set a west-to-east nonstop record in his Gamma, flying from Los Angeles to Floyd Bennett Field, New York, in 13 hours, 26 minutes, 15 seconds.

Lincoln Ellsworth donated the *Polar Star* to the Smithsonian in 1936.

THE N-1M (NORTHROP Model 1 Mockup) Flying Wing was a natural outgrowth of John K. "Jack" Northrop's lifelong concern for an aerodynamically clean design in which all unnecessary drag caused by protruding engine nacelles, fuselage, and vertical and horizontal tail surfaces would be eliminated. Developed in 1940, the N-1M was the first pure all-wing airplane produced in the United States.

For assistance in designing the aircraft, Northrop enlisted aerodynamicist Theodore von Karman, who was at the time director of the Guggenheim Aeronautical Laboratory at the California Institute of Technology, and von Karman's assistant, William R. Sears.

Nicknamed "Jeep," the N-1M emerged in July 1940 as a boomerang-shaped flying scale mockup built of wood and tubular steel with a wingspan of 11.6 m (38 ft.), a length of 5.2 m (17 ft.), and a height of 1.5 m (5 ft.). Pitch and roll were controlled by means of

DIMENSIONS:

WINGSPAN	11.6 M (38 FT.)
LENGTH	5.2 M (17 FT.)
HEIGHT	1.5 M (5 FT.)
WEIGHT	1,814 KG (4,000 LB.)

1. The museum's Northrop N-1M on display at the Udvar-Hazy Center. **2.** Jack Northrop, president of Northrop Aircraft, is standing next to the N-1M, with test pilot Moye Stephens sitting in the cockpit. **3.** The Northrop N-1M under construction on the Northrop factory floor near the end of its initial construction, circa mid-1940. **4.** The Northrop N-1M Flying Wing "Jeep" in flight.

elevons on the trailing edge of the wing, which served the function of both elevator and aileron. Instead of the conventional rudder there was a split flap device on the wingtips—these were originally drooped downward for what was thought to be better directional stability but were later straightened.

Controlled by rudder pedals, the split flaps, or "clamshells," could be opened to increase the angle of glide or reduce airspeed, and thus act as air brakes. The center of gravity, wing sweep, arrangement of control surfaces, and dihedral were adjustable on the ground. To decrease drag, the aircraft's two 65-hp Lycoming O-145 four-cylinder engines were buried within the fuselage. These lacked sufficient power to sustain lift and were replaced by two 120-hp six-cylinder 6AC264F2 air-cooled Franklin engines.

The N-1M made its first test flight on July 3, 1940, at Baker Dry Lake, California, with Vance Breese at the controls. During a high-speed taxi run, the aircraft hit a rough spot in the dry lake bed, bounced into the air, and accidentally became airborne for a few hundred yards. Later, during In the initial stages of flight testing, Breese

reported that the aircraft could fly no higher than 1.5 m (5 ft.) off the ground and that flight could only be sustained by maintaining a precise angle of attack. The problem was solved by making adjustments to the trailing edges of the elevons.

Another test pilot, Moye Stephens, reported that when attempting to move the N-1M about its vertical axis, the aircraft had a tendency to oscillate in what is called a Dutch roll. That is, the aircraft's wings alternately rose and fell, tracing a circular path in a plane that lies between the horizontal and the vertical. Stephens found that adjustments to the aircraft's configuration cleared up the problem. Although not known for sure how long, the N-1M was tested, it was for at least another six months.

From its inception, the N-1M was plagued by poor performance because it was both overweight and chronically underpowered. Despite these problems, the N-1M was successful enough to serve as the forerunner of more advanced flying wing aircraft like the N-9M, XB-35, YB-49, and ultimately the B-2 "stealth" bomber.

The N-1M was given to the Smithsonian in 1949, and restoration was finished in 1983.

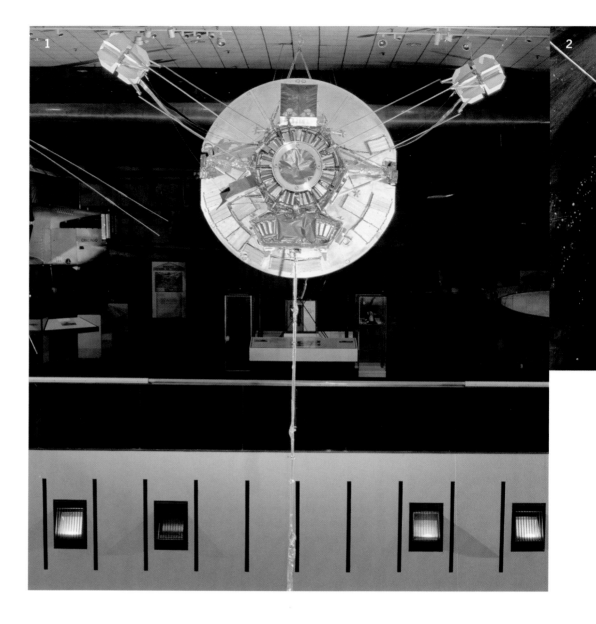

LAUNCHED IN 1972 and on its way out of the solar system, Pioneer 10 is one of the most distant space probes to leave Earth.

The objectives for Pioneer 10 (and 11) were to explore interplanetary space beyond Mars, to investigate the asteroid belt to learn if it posed a hazard to spacecraft bound for the outer planets, and to explore the environment of Jupiter. Later those objectives were extended to include the study of interplanetary space to extreme distances and the use of gravity assist, using Jupiter's gravitational field as a whip to propel the craft beyond Jupiter to the outermost planets.

Pioneer 10 was launched from the Kennedy Space Center on March 3, 1972, on a direct ascent trajectory, which means without first being placed in orbit around Earth. Just eleven hours after launch, the spacecraft passed the Moon and headed into interplanetary space. In July, Pioneer 10 entered the asteroid belt. Seven months later it emerged unscathed.

Pioneer 10 encountered Jupiter in early December 1973. Valuable data were returned while it was in transit. Especially significant were measurements of the intense magnetic

DIMENSIONS: ⎓⎓⎓⎓

ANTENNA DIAMETER   2.7 M (9 FT.)

LENGTH   2.9 M (9.5 FT.)

WEIGHT   258 KG (568 LB.)

3

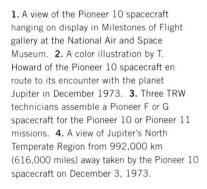

4

1. A view of the Pioneer 10 spacecraft hanging on display in Milestones of Flight gallery at the National Air and Space Museum. **2.** A color illustration by T. Howard of the Pioneer 10 spacecraft en route to its encounter with the planet Jupiter in December 1973. **3.** Three TRW technicians assemble a Pioneer F or G spacecraft for the Pioneer 10 or Pioneer 11 missions. **4.** A view of Jupiter's North Temperate Region from 992,000 km (616,000 miles) away taken by the Pioneer 10 spacecraft on December 3, 1973.

fields that surround Jupiter and their associated radiation belts, observations of the temperatures and the structure of Jupiter's upper atmosphere, and the return of color images of the planet, including Jupiter's red spot. Images from the scanning photopolarimeter were downloaded and stored on magnetic tapes, which were flown to Tucson for processing by the Optical Sciences Center at the University of Arizona. A 70-hour time lag was typically involved for full-image processing.

Since its Jupiter encounter, Pioneer 10 has continued its journey outward. It will eventually leave the solar system entirely. Because its power sources are long-lived radioisotope thermoelectric generators, Pioneer 10 continues to operate and send back data, including measurements of the solar magnetic field. In July 1981,

Pioneer 10 passed the 25 astronomical unit (AU) milestone, or 3.7 billion km (2.3 billion mi.) from the Sun. One AU equals the mean distance between the Sun and the Earth. Until February 1998, Pioneer 10 was the most distant space probe launched from Earth. But in that month, *Voyager 1* became the most distant space probe, at 10.4 billion km (6.5 billion mi.) from Earth. The two are headed in almost opposite directions away from the Sun.

NASA officially ended the Pioneer 10 mission on March 31, 1997. However, NASA still receives intermittent data from Pioneer 10.

The Pioneer 10 in the National Air and Space Museum is a full-scale model constructed from spare parts and is identical to the flight configuration. Pioneer 10 has been on display in the Milestones of Flight gallery since 1977.

1

THE PIPER J-3 earned its fame as a trainer and was so successful that the name "Cub" soon became synonymous with all light airplanes.

The story of the J-3 began in the late 1920s with C. Gilbert and Gordon Taylor, partners in the small Taylor Brothers Aircraft Company of Rochester, New York. Onetime barnstormers, the brothers were attempting to market a two-seat monoplane of their own design called the Chummy, when Gordon Taylor was killed in a crash. In 1929, Gilbert Taylor, who believed that there was a market for light planes, moved to Bradford, Pennsylvania, where community leaders were eager to promote new local industries. The Bradford Board of Commerce provided $500,000 to capitalize the new Taylor Company, which built five Chummys before the Depression ended construction.

One of the stockholders was oilman William T. Piper. Interested in aviation and believing that the Chummy was too expensive and inefficient, Piper offered to sponsor the development of a small plane to sell for half the Chummy's $3,985. The resulting aircraft, designated the E-2, was completed in late 1930 and fitted with a two-cylinder Brownbach

DIMENSIONS:

WINGSPAN	10.7 M (35 FT., 2.5 IN.)
LENGTH	6.83 M (22 FT., 4.5 IN.)
HEIGHT	1.9 M (6 FT., 8 IN.)
WEIGHT	309 KG (680 LB.)

1. NASM's Piper J-3 Cub hanging on display at the Udvar-Hazy Center. 2. A military version Piper L-4 suspended from a wire landing system known as Brodie gear.
3. A Piper J-3C-65-8 Cub parked at a store/filling station during the 1941 Louisiana Maneuvers. A cavalry unit is passing by the J-3. 4. An aerial view of Piper J-3 Cub.

Tiger Kitten engine. But testing revealed that the Tiger Kitten, which was rated at 20 hp, had too little power for the E-2. At full throttle, the small airplane was only able to rise a few feet into the air.

In 1930 the Taylor Company went bankrupt, but Piper bought the assets for $761 and retained Taylor as president. The aircraft received a Continental A-40 engine and became the Taylor E-2, resulting in sales of more than 300. Still not satisfied with the design and acting on owner suggestions, Piper hired Walter Jamouneau in 1935 to improve the aircraft. Jamouneau designed rounded wingtips, a curved rudder and fin, a true cabin, and a shaped and faired turtle deck aft of the wing. The Taylor J-2 (J for Jamouneau) emerged.

Not pleased with the new design, Taylor first fired Jamouneau but ran afoul of Piper and left the company in 1935 to found the Taylorcraft Airplane Company in Ohio. Piper rehired Jamouneau, and Taylor J-2s were produced until the factory burned to the ground in March 1937. Piper moved the company to an old silk factory in Lock Haven, Pennsylvania, and began production again in July. By November the aircraft name changed to Piper, and the first Piper J-2 Cub rolled off the line. In all, 1,547 E-2/J-2 Cubs were built from 1931 through 1938.

In 1938, Piper introduced the improved J-3 Cub. Powered by 40-hp Continental, Lycoming, or Franklin engines, the J-3 sold for $1,300. Engine horsepower was soon raised to 50 and reached 65 by 1940. Piper also standardized a color scheme to bright yellow with black trim.

The Civilian Pilot Training Program (CPTP), before the entry of the United States into World War II, spurred sales of the J-3. In 1940, 3,016 Cubs were built, and at the wartime peak a new J-3 emerged from the factory every twenty minutes. Seventy-five percent of all pilots in the CPTP were trained on Cubs, many going on to more advanced training in the military.

Cubs were also flown during the war as observation, liaison, and ambulance airplanes. Known variously as the L-4, 0-59, and NE-1, they rendered valuable service and were nicknamed "Grasshoppers." By 1947, when production ended, 19,888 Piper Cubs had been built. The J-3 is now finding an ever-increasing popularity among antique airplane owners, and brand-new Cubs are being constructed by airplane enthusiasts.

The National Air and Space Museum's Piper J-3 Cub was donated in 1977 and had logged 5,655 hours of flight time.

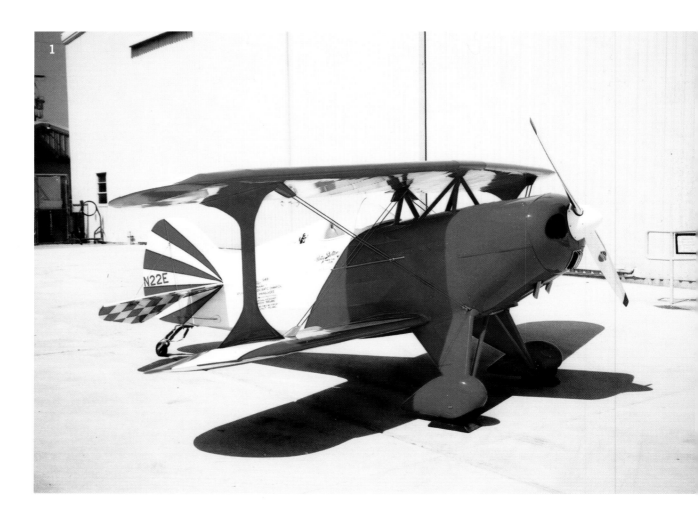

1

IN THE PITTS Special S-1C *Little Stinker,* aerobatic pilot Betty Skelton gained national and international recognition.

In 1945, Curtis Pitts set out to build a smaller aerobatic airplane that would climb, roll, and change attitude much more quickly. Instead of a large radial engine, Pitts built his aircraft around the new smaller and lighter horizontally opposed engines. Its swept wing allowed for access and center of gravity factors and made quicker snap rolls. The resulting Pitts Special S-1 was revolutionary because of its small size, light weight, short wingspan, and extreme agility. The S-1 was wrecked several years after its first flight.

The S-1C, with a slightly longer wing and fuselage and a Continental C-85-F5 engine, was built in 1946. Phil Quigley, Pitts's friend and test pilot, flew the bright red S-1C at air shows for a year. The airplane made such a good impression that Jess Bristow bought it and hired Quigley to fly it in his World Air Shows. He removed the original Continental C-85 engine and installed a C-90.

In August 1948, without having flown the aircraft, Betty Skelton bought the Pitts Special for $3,000. Skelton made several changes to it. For cross-country flight, her father constructed a small canopy that was easily and quickly removed for aerobatics. She replaced the original Aeromatic propeller with a fixed-pitch McCauley. She also mounted a ball-bank indicator upside down in the instrument panel just above the one used for normal flight, for control coordination in inverted flight.

Skelton won the 1949 Feminine International Aerobatic Championship held at the

1. A view of the Pitts S-1C Special *Little Stinker* following its restoration at the Paul E. Garber Preservation, Restoration, and Storage Facility. 2. Betty Skelton standing in the cockpit of her Pitts S-1C Special *Little Stinker*. 3. The cockpit of *Little Stinker*. 4. Betty Skelton at the controls of *Little Stinker*.

DIMENSIONS: ────────

WINGSPAN	4.9 M (16 FT., 10 IN.)
LENGTH	4.7 M (15 FT., 6 IN.)
HEIGHT	1.7 M (5 FT., 6 IN.)
WEIGHT	263 KG (580 LB.)

Miami All American Air Maneuvers in the S-1C. She also performed at a number of major air shows, including the International Air Pageant in London and the Royal Air Force Pageant in Belfast, Northern Ireland. In 1950, Skelton again won the Feminine International Aerobatic Championship in the Pitts, her third win overall.

A few Pitts Specials were built by Curtis Pitts and others in the 1950s, some for other female aerobatic pilots like Caro Bayley. But the airplane remained a minor type until the early 1960s, when interested amateurs, who remembered Skelton's remarkable aerobatic flying, convinced Pitts to produce a set of construction drawings

(at $125 per set). The popular homebuilt version of the S-1C had two ailerons, M-6 airfoils, and any engine from 85 hp up to 180 hp, the most popular being 125- to 150-hp Lycomings.

The reputation of the Pitts Specials is international. In 1966, Bob Herendeen became US National Aerobatic Champion in his S-1C. In the same year he competed in the World Championships in Moscow in that plane, arousing considerable interest in Europe.

In 1951, Betty Skelton retired from aerobatic competition, and in 1985 she donated *Little Stinker* to the National Air and Space Museum.

1

IN RESPONSE TO the Soviet's *Sputnik 1,* the Ranger program was created to gather information about the Moon.

Nine Rangers were launched using Atlas Agena B boosters from 1961 through 1965. Rushed development contributed to the failure of the first six Ranger missions. NASA took stock of the situation and redesigned the spacecraft to eliminate all but the television camera system. As a result, Rangers 7, 8, and 9 successfully transmitted more than 17,000 television pictures of the lunar surface.

Each Ranger spacecraft carried six cameras that differed in exposure times, fields of view, lenses, and scan rates. The camera system was divided into two channels, F (full) with a single wide-angle and narrow-angle camera, and P (partial) with two wide-angle and two narrow-angle cameras. Both channels had separate power supplies, timers, and transmitters. The F-channel returned images from as close as 5 m (16 ft.) above the Moon while the P-channel did so from as low as 600 m (1,969 ft.), just 0.2 seconds before impact.

DIMENSIONS: ⎯⎯

HEIGHT   3.1 M (10 FT., 3 IN.)
SPAN   4.6 M (15 FT.)
WEIGHT   366 KG (809 LB.)

4

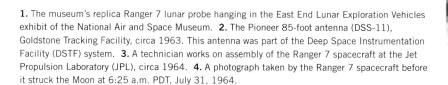

1. The museum's replica Ranger 7 lunar probe hanging in the East End Lunar Exploration Vehicles exhibit of the National Air and Space Museum. 2. The Pioneer 85-foot antenna (DSS-11), Goldstone Tracking Facility, circa 1963. This antenna was part of the Deep Space Instrumentation Facility (DSTF) system. 3. A technician works on assembly of the Ranger 7 spacecraft at the Jet Propulsion Laboratory (JPL), circa 1964. 4. A photograph taken by the Ranger 7 spacecraft before it struck the Moon at 6:25 a.m. PDT, July 31, 1964.

The photos enabled NASA to make highly detailed maps as well as three-dimensional representations of the lunar surface. The images were 1,000 times better than those available from existing Earth-based telescopes. Craters as small as 1 m (3.2 ft.) in diameter and geological evidence of volcanism were seen for the first time. During the last mission, the pictures were broadcast live on network television, enabling millions of viewers to witness a descent to the Moon.

The Deep Space Network that was created to track the progress of each Ranger flight provided invaluable scientific data on the dimensions and composition of the Moon. Because of Ranger, NASA gained experience in computing trajectories much more accurately by developing a sophisticated two-way Doppler tracking and communications system that measured the spacecraft's velocity between it and the tracking stations.

The Ranger program also helped NASA pioneer technologies in the design and construction of future robotic spacecraft intended for the exploration of deep space. In particular, attitude stabilization on three axes, onboard computers, and a steerable high-gain antenna was developed.

An important spin-off for the civilian sector involved digital computer image enhancement through techniques used to remove noise from Ranger's picture signals and to enhance contrasts in the photographs of the lunar surface. The techniques proved immediately useful in the enhancement of X-rays, providing doctors with much more accurate images of patients' skeletal structure. Ranger was transferred to the museum in 1977.

DURING WORLD WAR II, Republic P-47 pilots flew into battle with the roar of a 2,000-hp radial engine and the deadly flash of eight .50-caliber machine guns.

The first pilots to fly the P-47 in England in December 1942 greeted it with mixed emotions. Pilots of the 4th Fighter Group, Eighth Air Force, were accustomed to more nimble and lightweight fighters such as the Supermarine Spitfire (see page 180) and Hawker Hurricane. It weighed more than twice as much as the Spitfires many men had flown previously, so someone nicknamed the aircraft "Juggernaut," and "Jug" for short.

Early combat sorties revealed that the Thunderbolt could outdive all opposing fighters—a definite advantage in aerial combat. The P-47 could also absorb tremendous battle damage and continue to fly, and the eight machine guns gave it the greatest projectile throw-weight of any US fighter serving in World War II, except for the Northrop P-61 Black Widow night fighter.

However, initial operational experience revealed problems with the engine, radio, landing gear, range, and rate of climb. The first three difficulties were soon sorted out, but

WINGSPAN	12.7 M (41 FT., 9 IN.)
LENGTH	11 M (36 FT., 1 IN.)
HEIGHT	4.3 M (14 FT., 2 IN.)
WEIGHT	4,491 KG (9,900 LB.)

1. The museum's Republic P-47D Thunderbolt on display at the Udvar-Hazy with the Boeing B-29 Superfortress *Enola Gay* in background. **2.** Three Republic P-47N Thunderbolt fighters in stepped-formation flight over dense cloud layer. **3.** Lt. Quentin Anderson, 391st Fighter Squadron, 366th Fighter Group, in front of his crashed Republic P-47D Thunderbolt *Topsy* in France, August 3, 1944.

rate of climb was not dramatically improved until December 1943 when new broad-chord "paddle-blade" propellers were installed. Range limitations plagued the P-47 as long as it served in the European Theater. In the Pacific, the P-47N had a completely redesigned wing that held more fuel. This version could fly more than 3,220 km (2,000 mi.) and escort Boeing B-29 Superfortresses (see page 28) attacking the Japanese home islands.

The P-47 underwent many other modifications to improve its combat efficiency. The P-47D featured water injection to boost engine power, more powerful versions of the R-2800 engine, increased fuel capacity, and a bubble canopy for less-restricted visibility from the cockpit.

As a result of the US Lend-Lease policy, 247 Thunderbolts went to the British and 103 to the Soviet Union. The Brazilians flew them in the Italian Theater and in the Pacific, and the Mexican 201st Fighter Squadron used Thunderbolts in the Philippines.

Of the 15,683 P-47s built, about two-thirds reached overseas commands. A total of 5,222 were lost—1,723 in accidents not related to combat. The Thunderbolt flew more than half a million missions and dropped more than 132,000 tons of bombs. Thunderbolts were lost at the exceptionally low rate of 0.7 percent per mission, and their pilots achieved an aerial kill ratio of nearly 5 to 1. In the European Theater, P-47 pilots destroyed more than 7,000 enemy aircraft, more than half of them in air-to-air combat. They destroyed the remainder on very dangerous ground-attack missions. Between D Day and V-E Day pilots flying the Thunderbolt destroyed 86,000 railway cars, 9,000 locomotives, 6,000 armored fighting vehicles, and 68,000 trucks.

The Thunderbolt in the National Air and Space Museum is a P-47D-30-RA. It was donated to the Smithsonian in 1946, and it is now on display in the Steven F. Udvar-Hazy Center.

BURT RUTAN'S VARIEZE looked exotic and unusual, but it was easy to build, and it possessed good speed and range capabilities.

The VariEze's ("very easy") canard layout was visually striking, but its purpose went beyond aesthetics. Rutan designed the canard to stall before the wing, and this made the VariEze safer to fly.

The VariEzes built from plans did not use conventional ailerons mounted on the aft wing, rather elevons attached to the canard. Elevons are both elevators for pitch and ailerons for roll combined into one control. By using only elevons, the VariEze was easier to build.

Rutan also added winglets to both wingtips. Richard Whitcomb, an aerodynamics specialist at NASA Langley, had developed this technology during the 1950s. Winglets decrease drag to boost climb rate and cruise speed. On the VariEze, Rutan made the winglets perform double duty as vertical stabilizers and rudders to control yaw.

Rutan, his wife Carolyn, and friends worked for nearly four months to build the first VariEze in May 1975. It was powered by a modified Volkswagen engine. Three months later, the VariEze appeared at the annual Experimental Aircraft Association Convention and Fly-In held at Oshkosh, Wisconsin. The VariEze was so popular at Oshkosh that Rutan immediately began to work on designing a VariEze specifically for homebuilders. He and his team built the second VariEze during the winter of 1975 with a heavier, more reliable Continental O-200 engine and added 4.5 kg (10 lb.) of ballast to the nose. Every dimension changed, and the wing area increased from 5.3 sq. m (59 sq. ft.) to 6 sq. m (67 sq. ft.). Rutan modified the canard as well and improved the airfoil.

DIMENSIONS:

WINGSPAN	6.8 M (22 FT., 2.5 IN.)
LENGTH	4.3 M (14 FT., 2 IN.)
HEIGHT	1.5 M (59 IN.)
WEIGHT	263 KG (585 LB.)

1. The museum's Rutan VariEze on display at the Udvar-Hazy Center with its nose gear retracted.
2. Two Rutan Model 33 VariEze aircraft on display at the International Experimental Aircraft Association (EAA) Fly-In Convention, at Oshkosh, Wisconsin. **3.** Grand view of two of the uniquely designed canard VariEzes.

Rutan began publishing plans in July 1976, and the first homebuilt VariEzes began to fly less than a year later. At that time, it was rare indeed for builders to go so quickly from a set of plans to a flying, high-performance airplane. The key to such rapid construction lay in Rutan's choice of composite construction. To build the composite fiberglass/foam/fiberglass sandwich structure, the builder used a hot-wire cutter to slice the foam core to shape, then cover the core with epoxy resin and fiberglass cloth cut precisely to maintain strength at minimum weight.

When the homebuilders began to fly their VariEzes, two problems immediately surfaced that involved weight and control of the airplane. Rutan discovered that the average finished airplane was 14 to 23 kg (30 to 50 lb.) heavier than specified in the plans. Some builders had disregarded Rutan's fundamental design philosophy of a simple, lightweight aircraft equipped to operate in daylight and clear weather and had added heavy instrument packages to allow them to fly at night and in bad weather.

This equipment required more complex and heavier electrical systems. Some VariEzes weighed so much that they could not safely carry a passenger.

The control issue demanded immediate action. Considerable precision was required to build the correct incidence and twist into the aft wing. Errors in wing incidence and twist were overpowering the elevons mounted on the canard. Rutan decided to redesign the rear wing to incorporate ailerons but to keep the elevator control in the canard. He and his team completed this task in about two weeks and rushed the changes to anxious builders.

The VariEze became extremely popular, and Rutan eventually sold about 3,000 sets of plans before he stopped selling them in 1985 in order to pursue other business interests. Two hundred VariEzes were flying by 1980.

The VariEze in the National Air and Space Museum had clocked 700 flying hours and was used by NASA scientists to study stall/spin accidents before it was delivered to the museum in 1986.

ON DECEMBER 23, 1986, the Rutan *Voyager* completed the first nonstop and nonrefueled flight around the world.

Made of lightweight composite materials in 98 percent of its structure, *Voyager* was designed for maximum fuel efficiency. Most of the airplane is made from a 0.635-cm (0.25-in.) sandwich of paper honeycomb and graphite fiber, carefully molded and cured in an oven. The entire airframe was constructed without using metal and weighs only 425 kg (939 lb.). *Voyager* took more than 22,000 work hours and more than 18 months to construct.

The long, thin main wing was so flexible that the wingtip deflected upward 0.9 to 1.5 m (3 to 5 ft.) while the aircraft was in flight. The purpose of the winglets was to raise the fuel vent from the three outboard wing tanks high enough to keep the fuel from draining out onto the ground. The cabin and cockpit were side by side within the fuselage.

Voyager was virtually a flying fuel tank. It had eight storage tanks on each side of the airplane and a fuel tank in the center, for a total of seventeen tanks. The pilot shifted fuel from tank to tank during the flight to keep the airplane in balance. The 3,181 kg (7,011 lb.) of fuel aboard at takeoff amounted to 72.3 percent of its gross weight. At the end of the flight only 48 kg (106 lb.) of fuel remained.

Two engines, one at each end of the fuselage, powered the aircraft. The highly efficient 110-hp, liquid-cooled rear engine, a Teledyne Continental IOL-200, ran during the entire flight except for four minutes when a fuel problem caused a temporary shutdown. The 130-hp, air-cooled front engine, a Teledyne Continental O-240, was used for a total of 70 hours and 8 minutes during the initial, heavyweight stage of the flight, and also while climbing over weather and at other critical times.

WINGSPAN	33.8 M (110 FT., 8 IN.)
LENGTH	8.9 M (29 FT., 2 IN.)
HEIGHT	3.1 M (10 FT., 3 IN.)
WEIGHT	1,020 KG (2,250 LB.)

1. A left-front view of Rutan Model 76 *Voyager* hanging on exhibit in the Independence Avenue lobby of the National Air and Space Museum. **2.** The Rutan *Voyager* in flight over Edwards Air Force Base, California. **3.** The Rutan *Voyager* aircraft in flight. **4.** An interior view of the Rutan *Voyager* cockpit's forward instrument panel.

Voyager was equipped with Hartzell constant-speed, variable-pitch aluminum propellers that proved to be a critical factor in stretching the aircraft's range enough to bring it home. These propellers were designed, built, and delivered in only seventeen days after one of the original propellers failed.

Voyager's takeoff roll lasted 2 minutes and 6 seconds with less than 244 m (800 ft.) of the 4,572-m (15,000-ft.) runway left at liftoff. Both winglets were damaged during takeoff when the wingtips dragged along the runway. Soon after takeoff, an attempt was made to deliberately dislodge the damaged winglets over Edwards Air Force Base so that if structural damage occurred to the aircraft's fuel tanks, it could land as soon as possible. To dislodge the winglets, the crew increased flight speed and maneuvered the airplane to build up side forces sufficiently high to break them free. About 8 km (5 mi.) from the base, the right winglet was dislodged and fell into someone's yard. The other winglet was also successfully dislodged, but was never found.

Following a route determined by weather, wind, and geography, *Voyager* flew an official distance of 42,212 km (24,986 mi.) at an official speed of 186 km/h (116 mph), in an elapsed time of 216 hours, 3 minutes, and 44 seconds. Flying *Voyager,* Dick Rutan and Jeana Yeager established eight absolute and world-class records. On the second day of the flight over the Pacific Ocean, they were guided by meteorologists through the edges of Typhoon Marge for a slingshot effect from tailwinds, increasing ground speed to 130 knots.

Throughout the flight, the physical and mental capabilities of the pilots were continually tested by mechanical and severe weather problems, as well as by the cramped quarters. The pilot in the cockpit flew the airplane, navigated, maintained ground communication, and transferred fuel to balance the airplane. The other pilot rested, managed the logistics support tasks of the flight, or provided navigation and flight-monitoring assistance. As a result of the record-breaking flight, *Voyager* earned the Collier Trophy, aviation's most prestigious award.

Donated in 1987, *Voyager* now hangs in the south lobby of the National Air and Space Museum.

2

ON MAY 21, 1927, Charles A. Lindbergh completed the first solo nonstop transatlantic flight in history, flying his Ryan NYP *Spirit of St. Louis* 5,810 km (3,610 mi.) between Roosevelt Field, Long Island, New York, and Paris in 33 hours and 30 minutes.

Lindbergh studied aeronautics at the Nebraska Aircraft Corporation and was a barnstormer before he enrolled as a flying cadet in the US Army Air Service. He won his reserve commission and served as a civilian airmail pilot, flying the route between St. Louis and Chicago.

Early in 1927 he obtained the backing of several St. Louis men to compete for the $25,000 prize offered by Raymond Orteig in 1919 for the first nonstop flight between New York City and Paris. In February of that year, Lindbergh placed an order with Ryan Airlines in San Diego for a modified M-2.

Development began based on a standard Ryan M-2, with Donald A. Hall as principal designer. The M-2 was a basic high-wing, strut-braced monoplane. Lindbergh had the wingspan increased by 3 m (10 ft.), and the fuselage and wing cellule were redesigned to accommodate the greater fuel load. Plywood was fitted along the leading edge of the wings. The fuselage was lengthened .6 m (2 ft.). The cockpit was moved farther to the rear for safety, and the engine was moved forward for balance, permitting the fuel tanks to be installed at the center of gravity. The pilot looked forward through a periscope or by turning the aircraft to look out of a side window. A Wright Whirlwind J-5C engine supplied the power.

DIMENSIONS:

WINGSPAN	14.02 M (46 FT.)
LENGTH	8.41 M (27 FT., 7 IN.)
HEIGHT	2.99 M (9 FT., 10 IN.)
WEIGHT	975 KG (2,150 LB.)

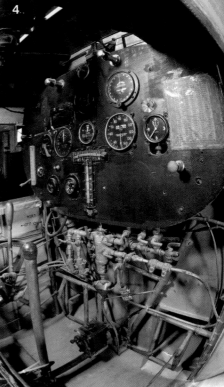

1. The *Spirit of St. Louis* hanging on public display in the Milestones of Flight hall at the National Air and Space Museum. **2.** Charles A. Lindbergh posed in front of the *Spirit of St. Louis* in San Diego, California, prior to flight to New York. **3.** The *Spirit of St. Louis,* piloted by Charles A. Lindbergh, taking off from San Diego on April 10, 1927. **4.** The cockpit of *Spirit of St. Louis,* showing forward instrument panel, stick, rudder pedals, and fuel control system.

Late in April 1927, the *Spirit of St. Louis* was ready. It was painted silver and carried registration number N-X-211, which, with all other lettering on the plane, was painted in black. After several test flights, Lindbergh flew from San Diego to New York on May 10 to 12, stopping once in St. Louis. His flight time of 21 hours and 40 minutes set a new transcontinental record.

After waiting several days in New York for favorable weather, Lindbergh took off for Paris alone on the morning of May 20, 1927. Thirty-three hours, 30 minutes, and 5,810 km (3,610 mi.) later he landed safely at Le Bourget Field near Paris, where he was greeted by a crowd of 100,000.

Lindbergh and the *Spirit of St. Louis* returned to the United States aboard the USS *Memphis* on June 11. He received tumultuous welcomes in Washington, DC, and New York City. From July 20 to October 23 of that year he took the famous plane on a tour of the United States. Then, on December 13, he began a tour that took him to Mexico City, Central America, Colombia, Venezuela, Puerto Rico, and Havana. Beginning in Mexico City, flags of the countries he visited were painted on both sides of the cowling.

The aftermath of the flight was the "Lindbergh boom" in aviation: aircraft industry stocks rose in value and interest in flying skyrocketed. Lindbergh's subsequent tours in the *Spirit of St. Louis* demonstrated the potential of the airplane as a safe and reliable mode of transportation.

On April 30, 1928, the *Spirit of St. Louis* made its final flight from St. Louis to Washington, DC, where Lindbergh presented the aircraft to the Smithsonian.

THE KEY TO the Apollo space program was the power and reliability of the F-1–powered Saturn V launch vehicle, which carried the astronauts into Earth orbit and beyond to the Moon.

One of the most powerful single liquid-fuel engines ever utilized in rockets, the F-1 was developed and built by the Rocketdyne Division of North American Aviation (later Rockwell International), for NASA. Capable of 1.5-million-pound thrust, five F-1 engines (at 7.5 million pounds total) powered the first stage of the three-stage Saturn V launch vehicle.

Saturn V was used for twelve Apollo flights from 1967 to 1972, the last seven of which (Apollo 11 through Apollo 17) were manned landing missions to the Moon. It was also used to put the Skylab (see page 172) orbital workshop into Earth orbit in May 1973.

In the late 1950s medium-range ballistic missiles and intercontinental ballistic missiles developed by the military had been used to lift manned spacecraft into Earth orbit, but not into lunar orbit. A team at the Army Ballistic Missile Agency, led by Wernher von Braun, sought an engine capable of sending manned spacecraft into lunar orbit. Much of this team was transferred to NASA following its creation late in 1958. NASA then awarded Rocketdyne a contract in early 1959 to develop the F-1. This engine used liquid oxygen (lox) and RP-1, a kerosene, and was developed by Rocketdyne under NASA's guidelines.

Parallel to this project, work proceeded on the family of Saturn launch vehicles. The final result was the three-stage Saturn V, which used five F-1s to power the first stage, five

LENGTH	5.4 M (18 FT.)
DIAMETER	3.7 M (12 FT.)
WEIGHT	8,319 KG (18,340 LB.)

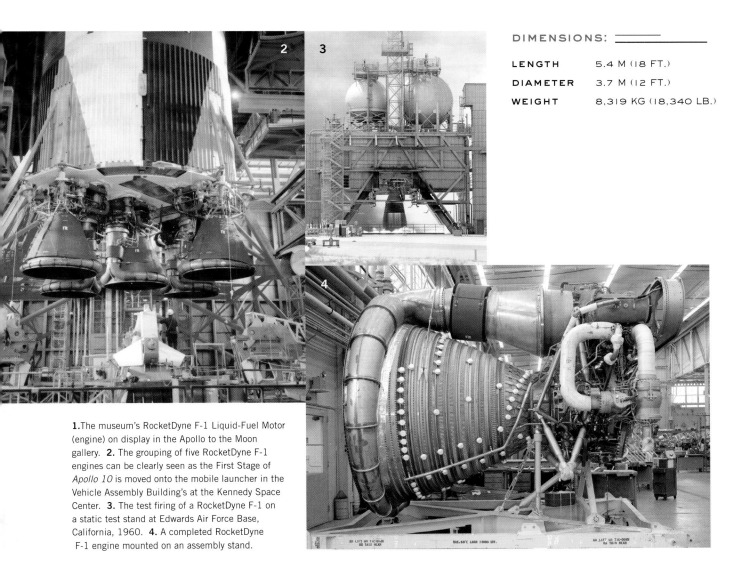

1.The museum's RocketDyne F-1 Liquid-Fuel Motor (engine) on display in the Apollo to the Moon gallery. **2.** The grouping of five RocketDyne F-1 engines can be clearly seen as the First Stage of *Apollo 10* is moved onto the mobile launcher in the Vehicle Assembly Building's at the Kennedy Space Center. **3.** The test firing of a RocketDyne F-1 on a static test stand at Edwards Air Force Base, California, 1960. **4.** A completed RocketDyne F-1 engine mounted on an assembly stand.

Rocketdyne J-2s for the second stage, and one J-2 for the third stage.

After tests with unmanned Apollo spacecraft (*Apollo 4* and *Apollo 6*) into Earth orbit, the Saturn V lifted the *Apollo 8* crew into lunar orbit. This was the first time astronauts had escaped Earth's gravitational pull.

Most notably, the Saturn V was used for the Apollo 11 mission in July 1969, which sent two astronauts to the surface of the Moon. Remarkably, the F-1s worked perfectly every time the Saturn V went into space. It was transferred from NASA in 1970.

THE MOST SUCCESSFUL aircraft in air racing history, *Nemesis* dominated its competition and won forty-five of forty-eight contests from 1991 until its retirement in 1999.

Flown by pilot and designer Jon Sharp, it won nine consecutive Reno Gold National Championships and sixteen world speed records for its class, including the 3-km (1.8-mi.) mark of 466.84 km/h (290.08 mph) and the 15-km (9.3-mi.) mark of 454.77 km/h (282.58 mph) set in 1998. *Nemesis* was the International Formula 1 points champion for 1994, 1995, 1996, 1997, and 1998, at an average speed 393.77 km/h (244.68 mph). In 1993, 1996, and 1998 it won the Louis Blèriot Medal of the Fédération Aéronautique International for the greatest achievement in speed. In 1993, 1994, 1995, and 1999, *Nemesis* won the Pulitzer Trophy for air-racing speed records.

Built in 1991, *Nemesis* is a small mid-wing, single-seat tractor monoplane with fixed landing gear. It is constructed of a pressure-molded graphite epoxy foam core sandwich and is powered by a single Continental O-200, 100-hp air-cooled engine.

Nemesis had numerous firsts in the Formula 1 class. It was the first to be built entirely of tooled, pressure-molded, carbon-reinforced plastics, the first with a carbon fiber roll over structure, and the first to use a custom-designed natural laminar flow wing. It is the first to be entirely computer lofted.

Nemesis also pioneered the use of a side stick with a full ball bearing control system as well as an onboard data-acquisition system and a titanium firewall. It is the first racer with a center section containing the fuselage with an integral wing, landing gear, and wheel pants. These innovations earned the team the 1991 George Owl Trophy, awarded for design excellence.

DIMENSIONS: ━━━━━

WINGSPAN	6.25 M (20 FT., 6 IN.)
LENGTH	5.64 M (18 FT., 6 IN.)
WEIGHT	236 KG (520 LB.)
TOP SPEED	467 KM/H (290 MPH)

1. The Sharp *Nemesis* on display at the Udvar-Hazy Center. **2.** The Sharp *Nemesis* on display at the International Experimental Aircraft Association (EAA) Fly-In Convention, Oshkosh, Wisconsin, in August 2000 before it was officially donated to the museum.

With Jon Sharp at the controls, *Nemesis* won the first competition in which it was entered, the Gold Race at Reno, Nevada. It was the first aircraft to do so since the inaugural 1947 event. Through 1996, *Nemesis* won an astonishing thirty consecutive races while setting two national qualifying records. *Nemesis* also set new world speed records for Group 1 aircraft on a 3-km (1.86-mi.) course in 1993 at 446.21 km/h (277.26 mph) and 456.65 km/h (283.75 mph) three years later.

Nemesis was retired in 1999 and donated to the National Air and Space Museum. Fittingly, it is on display at the Steven F. Udvar-Hazy Center next to other prominent air racers.

SKYLAB WAS A space station launched into Earth orbit by the United States in May 1973. Skylab had four major elements: the orbital workshop, which housed living quarters and some experiment apparatus; the airlock module, with a hatch for extravehicular activity outside the station; the multiple docking adapter, with docking ports for the crew vehicle as well as a control center for the solar observatory and several other experiments; and a solar observatory called the Apollo telescope mount.

The orbital workshop was the largest component of Skylab. The crews lived and did most of their scientific research in the workshop. The outer surface includes a gold coating to reflect the Sun's heat and help control interior temperature. Under the workshop are twenty-three spherical containers for gaseous nitrogen used in the thrust attitude control system and pneumatics. A radiator for the life-support systems, refrigerators, and freezers is mounted below the spheres.

Power was supplied via large solar arrays. Twin solar array wing panels were folded against the orbital workshop for launch, one on each side. When Skylab reached orbit, the arrays were to be extended, exposing the solar cells to the Sun to produce 12 kilowatts of power.

DIMENSIONS:

HEIGHT	15 M (48 FT.)
DIAMETER	6.5 M (22 FT.)
WEIGHT	35,400 KG (78,000 LB.)

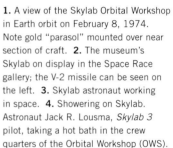

1. A view of the Skylab Orbital Workshop in Earth orbit on February 8, 1974. Note gold "parasol" mounted over near section of craft. **2.** The museum's Skylab on display in the Space Race gallery; the V-2 missile can be seen on the left. **3.** Skylab astronaut working in space. **4.** Showering on Skylab. Astronaut Jack R. Lousma, *Skylab 3* pilot, taking a hot bath in the crew quarters of the Orbital Workshop (OWS).

During the launch, the micrometeoroid shield accidentally deployed too soon, jamming one solar array wing and damaging the other so badly that both the wing and the shield were torn loose from the workshop. As a result, Skylab had only one solar array wing once in orbit. With clever engineering and improvisation, Skylab operations were completed despite the reduction in power.

The micrometeoroid shield that was torn loose during launch had another purpose: shading Skylab to control the temperature inside. The first crew deployed a temporary parasol-like sunshield. The second crew installed a larger sail-like sunshield to replace the temporary one.

The living area aboard Skylab was about the same as a small house and was reasonably spacious and comfortable. When in orbit, the Skylab space station was 36 m (118 ft.) long. With a docked Apollo command and service module, it weighed about 90,600 kg (100 tons).

Skylab was equipped with instruments and experiments to observe and study natural resources and the environment on Earth and to observe and study high-energy activity on the Sun. The missions were also intended to study the effects of weightlessness on the human body and assess crew adaptation to long-duration spaceflight, study materials processing in microgravity, and to perform experiments submitted by students for a "Classroom in Space" program.

Skylab was designed for visits as long as one to three months. The crews appreciated the large window for viewing Earth, the galley and wardroom with a table for group meals, the private sleeping quarters, and a shower custom-designed for use in weightlessness.

The Skylab missions obtained vast amounts of scientific data, and they demonstrated that people could live and work productively in space for months at a time. Skylab was intended to be temporary. Abandoned in 1974, it was largely destroyed by reentry in Earth's atmosphere in July 1979.

Two complete Skylab space stations were manufactured, and NASA donated the backup Skylab to the National Air and Space Museum, where it has been on display since 1976.

THE FIRST SPACE shuttle, *Enterprise,* is a test vehicle equipped to operate in the atmosphere and never flew in space.

In the post-Apollo era, the space shuttle was intended to make access to space routine and less expensive. To meet these goals, it had to be reusable and economical. It was to have a variety of purposes, including satellite delivery and retrieval, orbital servicing, round-trip service for science instruments, and laboratory research in space. The Department of Defense was also interested in using the shuttle to launch reconnaissance and other military satellites.

For the multipurpose space shuttle, the roles of astronauts expanded to include mission specialists who are responsible for spacecraft systems and operations other than piloting, as well as scientific research in space. For many missions, guest astronauts—called payload specialists—who are not part of the NASA astronaut corps trained with the crew. Payload specialists come from universities, research centers, government agencies, businesses, and other nations.

The space shuttle consists of an orbiter, an external propellant tank, and two solid rocket boosters. Only the orbiter goes into space. Designed to operate on land, in the atmosphere, and in space, the orbiter combines features of a rocket, an aircraft, and a glider. The space shuttle's liquid-propellant main engines and solid rocket motors are the first ever designed for reuse.

2 3

1. The museum's space shuttle *Enterprise* on display, at night, in the space hangar of the Udvar-Hazy Center.
2. The space shuttle orbiter *Enterprise* mated to the Boeing 747-123 Shuttle Carrier Aircraft (SCA) closely followed by a NASA Northrop T-38 chase plane.
3. The *Enterprise* nearing touchdown at the end of the first Orbiter Approach and Landing Test (ALT) flight, August 12, 1977, at the NASA Dryden Flight Research Center, California.
4. The crew insignia for the space shuttle (OV-101) Orbiter Approach and Landing Tests; the crewmen were Gordon Fullerton, Fred Haise, Joseph Henry Engle, and Richard Truly.

4

DIMENSIONS:

WINGSPAN	23.77 M (78 FT.)
LENGTH	37.17 M (122 FT.)
HEIGHT	17 M (57 FT.)
WEIGHT	68,586 KG (151,205 LB.)

The space shuttle is the only US vehicle currently used for human spaceflight. Three shuttle orbiters are in service: *Discovery, Atlantis,* and *Endeavour.* Two orbiters—*Challenger* and *Columbia*—have been destroyed by accidents in which their two crews have died. The shuttles usually fly two to nine missions a year.

The main role of the test vehicle *Enterprise* (OV-101, for Orbital Vehicle 101) was to check the shuttle's flight characteristics and performance. NASA conducted the approach and landing test program from February through November 1977 at the Dryden Flight Research Center in California. Ground tests included taxi tests with *Enterprise* mounted atop the Boeing 747 shuttle carrier aircraft to determine loads, control characteristics, steering, and braking of the mated vehicles. Five unmanned flights of *Enterprise* attached to the 747 were conducted to assess

structural integrity and performance of the mated craft in flight. Still attached to the 747, three manned captive flights followed with an astronaut crew. Finally, two astronaut crews took turns piloting the 68,039-kg (150,000-lb.) shuttle to five free-flight landings at Edwards Air Force Base under conditions simulating a return from space.

Then *Enterprise* was used for vibration tests at Marshall Space Flight Center in Alabama and for launch complex fit checks at Kennedy Space Center in Florida and Vandenberg Air Force Base in California. In 1983 it appeared in the Paris air show and other sites in Europe, and it was a featured attraction at the 1984 World's Fair in New Orleans.

In 1985, NASA transferred *Enterprise* to the Smithsonian. It is now on display at the Steven F. Udvar-Hazy Center.

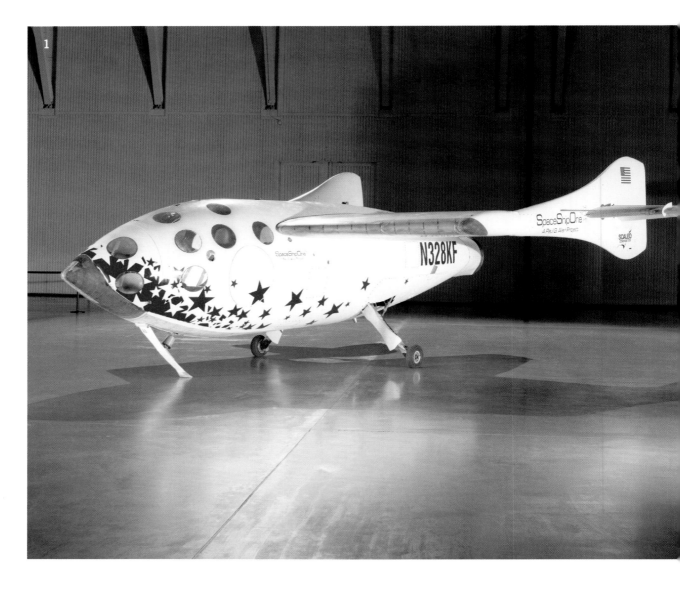

IN 2004, THE SpaceShipOne team achieved a major milestone in space history when it designed, built, and flew the first piloted flights in a privately developed spacecraft. SpaceShipOne was a collaboration between legendary aircraft designer Burt Rutan and investor and philanthropist Paul G. Allen. Rutan is well-known as the highly innovative designer of personal and special purpose aircraft such as the popular VariEze homebuilt series, the famous *Voyager,* which completed the first nonstop, nonrefueled around the world flight, and the *Virgin Atlantic Global Flyer,* which accomplished the same feat with but a single pilot. One of the cofounders of Microsoft, Allen is captivated by new ideas that solve important problems and improve people's lives and has invested extensively to achieve this goal.

Early in the first decade of the 21st century, Rutan and Allen focused their collective resources on solving the problem of private spaceflight in the hope of opening space travel to ordinary people. Inspired by this shared vision of space travel, the team received the first launch license for a privately developed spacecraft they called SpaceShipOne.

In appearance and operation, SpaceShipOne is unlike any spacecraft built before it. With a fuselage faintly resembling the bullet-shaped Bell X-1 rocket plane, SpaceShipOne has distinctive swept wings with tail fins. For its initial ascent it is tucked under a graceful long-winged aircraft also designed and built by Rutan's Scaled Composites company called

DIMENSIONS:

WINGSPAN	5 M (16 FT., 4 IN.)
LENGTH	8.5 M (27 FT., 11 IN.)
HEIGHT	2.7 M (8 FT., 11 IN.)
WEIGHT	1,361 KG (3,000 LB.), EMPTY

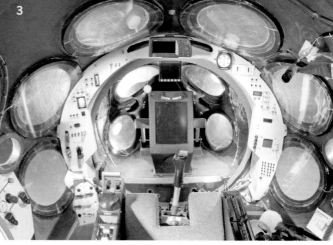

1. SpaceShipOne arrives at NASA's Udvar-Hazy Center. 2. "Captive Carry." The White Knight turbojet launch aircraft with *SpaceShipOne* mounted beneath in flight test over the Mohave Desert, California. 3. The cockpit of SpaceShipOne. 4. SpaceShipOne in flight as it glides down for approach to the Mojave airport.

White Knight. At 50,000 feet SpaceShipOne is released and the pilot ignites a hybrid rocket motor that produces 17,000 pounds of thrust from a mixture of solid rubber fuel burned with liquid nitrous oxide. The engine burns for 80 seconds, enough to reach Mach 3 and 180,000 feet. The vehicle then coasts to an altitude of more than 328,000 feet.

Once in space and on the way to apogee, the pilot reconfigures the craft. The twin tails and about a third of the wing tilt up, and SpaceShipOne becomes a stable "shuttlecock" for reentry into the atmosphere. While the pilot enjoys the view, the vehicle begins its descent. After the reentry deceleration, the pilot lowers the wings and tail back into position for atmospheric flight and glides to a runway landing.

Pilot Mike Melvill took SpaceShipOne to 62 miles (100 kilometers) on June 21, 2004, and to 64 miles on September 29. Brian Binnie flew it to 70 miles on October 4, 2004. The October record-setting flight was made to coincide with the 47th anniversary of the flight of *Sputnik 1,* the world's first artificial satellite. For the last two flights, Burt Rutan, Paul Allen, and the SpaceShipOne team won the $10 million Ansari X-Prize. They also received the 2005 National Air and Space Museum Trophy, an *Aviation Week* Laurel Award, and the 2004 Collier Trophy, as well as many other accolades.

Innovations from the SpaceShipOne program are making breakthroughs for safety that will allow commercial flight operations for public access to space. Among the achievements are the designing of a small, lightweight craft for both subsonic and supersonic flight; mastering the aerodynamics and flight controls for those different regimes; using the feathered wing as a brake to provide a "care-free" reentry; innovating with composite materials and a hybrid rocket engine; keeping the vehicle design simple, robust, and economical; and navigating the regulatory challenges for commercial, piloted launches into space. They demonstrate how essential teamwork is to successful spaceflight.

The creators of SpaceShipOne are committed to a commercial venture making spaceflight safe and accessible to the public, and their ultimate goal is to open space to tourism. It is their dream that someday, anyone might be able to buy a ride into space, experience weightlessness, and return safely home. It came to the museum in 2005.

FLOWN BY SOME of the most famous air heroes of the war, the fast and rugged Spad XIII was among the most successful fighters of the First World War.

Louis Béchereau, a talented designer who stayed with Spad after the reorganization of the company, was responsible for the successful series of Spad fighters in World War I. In 1915, he recognized the design limits of the then popular air-cooled rotary engines that powered most French aircraft. A fixed, water-cooled engine, capable of generating at least 150 horsepower, was in his view the power plant of the future. At this time, just such an engine was undergoing trials. It was a new Hispano-Suiza V-8, designed by that company's founder and chief designer, a Swiss-born engineer named Marc Birkigt. The engine was a great step forward in that it had a much better power-to-weight ratio than previous water-cooled designs, as well as many other modern features. The new Hispano-Suiza engine was just what Béchereau was calling for. He designed his next airframe around the advanced Hispano-Suiza motor; the result was the famous Spad VII. By the fall of 1916, Spad VIIs were entering French squadrons and were well received. More than five-and-a-half thousand Spad VIIs were built.

Manufactured by the Société pour l'Aviation et ses Dérives, the Spad XIII was a larger, improved version of the earlier Spad VII with, among other improvements, two fixed, forward-firing Vickers machine guns and a more powerful 200-hp Hispano-Suiza 8Ba engine. (Later Spad XIIIs had 220- and 235-hp Hispano-Suiza V-8 engines.) The prototype Spad XIII made its first flight on April 4, 1917, and by the end of the following month, production aircraft were arriving at the front. They were particularly noted for their robust

WINGSPAN	8.2 M (26 FT., 11 IN.)
LENGTH	6.3 M (20 FT., 8 IN.)
HEIGHT	2.4 M (7 FT., 11 IN.)
WEIGHT	823 KG (1,815 LB.), EMPTY

1. NASM's restored SPAD XIII *Smith IV,* July 9, 1986, at the Paul E. Garber Facility. 2. The cockpit of the museum's SPAD XIII. Note the three Maltese crosses have been painted to the left and below the replacement windscreen. 3. The pilot of the museum's Spad XIII, Lt. Arthur Raymond "Ray" Brooks, poses beside his Spad XIII, of the 22nd Aero Squadron in mid-1918.

construction and ability to dive at high speed. It was one of the best dogfighting airplanes of World War I.

The Spad XIII was produced and deployed in great numbers. By the end of 1918 the parent company and eight other French manufacturers had built 8,472. Almost every French fighter squadron was equipped with them by the end of the war, as well as the American units that were part of the American Expeditionary Force. Spads were also used by the British, Italians, Belgians, and Russians.

Surprisingly, given the large number built, only four Spad XIIIs remain. The one in the National Air and Space Museum, nicknamed *Smith IV,* was flown by Lt. A. Raymond Brooks, US Army Air Service, who named it after the college his sweetheart and future wife had attended. Brooks achieved one of his six personal aerial victories in *Smith IV.* Other victories were scored with *Smith IV* while flown by other pilots.

After the war, *Smith IV* toured the United States as part of Liberty Bond drives. It was then transferred to the Smithsonian in 1919. By the 1980s, *Smith IV*'s fabric was rotting and tattered, its tires missing, and the airplane was in a general state of disrepair. It was fully restored between 1984 and 1986 and placed back on public display in the museum's First World War gallery.

1

WITH A WATCHFUL eye on political developments in Germany in the 1930s, the British government issued specifications for a fighter aircraft with eight instead of the usual four machine guns. The Supermarine works at Woolston, England, was ready with the design for the Supermarine Type 300. It surpassed Air Ministry requirements and was accepted.

On March 5, 1936, Spitfire prototype took off from Eastleigh Airfield, Southhampton, on its maiden flight. After official trials an order for 310 planes was placed by the Air Ministry. When war with Germany was declared, 400 Spitfires were already in service and 2,160 were on order.

Designed by Reginald Mitchell, the Spitfire was an all-metal cantilever monoplane. The shape of the wing, its most distinguishing characteristic, was elliptical, which greatly reduced drag and increased speed. Even while the first deliveries were being made, improvements were being introduced. A metal two-blade controllable pitch propeller replaced the two-blade fixed pitch mahagony airscrew, increasing speed. A tailwheel replaced the tail skid. Bulletproof windshields were installed in Spits already in service and as they came off the production line.

The Spitfire was easy to handle. It became airborne quickly, and, once in the air, its maneuverability was outstanding. The combination of its speed and firepower made the Spit a deadly machine. Its eight machine guns concentrated a hail of bullets capable of tearing enemy planes 274.32 m (300 yd.) away.

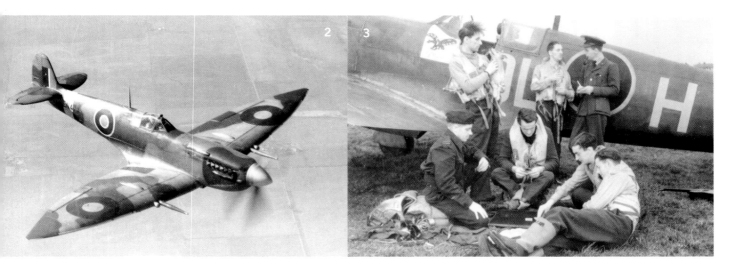

1. NASM's Supermarine Spitfire Mk.VII on display in the World War II Aviation gallery. 2. The museum's Spitfire Mk.VII in flight during a wartime test flight over Ohio. 3. Members of 91 Squadron of the Royal Air Force pose informally beside Supermarine Spitfire Mk.VII in late 1942. 4. A wartime photograph of the museum's Mk.VII Spitfire taken on February 24, 1944, at Wright Field, Ohio.

DIMENSIONS:

WINGSPAN	17.3 M (40 FT., 2 IN.)
LENGTH	9 M (29 FT., 11 IN.)
HEIGHT	3.58 M (11 FT., 5 IN.)
WEIGHT	2,670 KG (5,887 LB.), EMPTY

Because the Spitfire was designed principally as a home defense interceptor fighter, its range was limited. But in 1943 this was increased by adding external fuel tanks that could be jettisoned. This modification enabled the Spitfire to escort bombers to and from targets across the English Channel.

As needs arose, variations on the Spitfire were developed, including photo-reconnaissance versions, high-altitude versions to take on the Messerschmitt Bf 109s (see page 130), and low-altitude versions to meet the Focke-Wulf Fw 190s (see page 68). They were also employed in sea-air rescue operations. A Spitfire could quickly reach a downed pilot and drop a dinghy and emergency supplies, often saving him from the cold waters of the English Channel.

The Spitfire was in service with many different groups and on many different fronts. Belgians, French, Poles, Czechs, Americans, and British Commonwealth countries all used the fighter. The Eagle Squadron was one of the best known of these foreign units. Composed of American volunteers, the first Eagle Squadron was officially formed on October 19, 1940. When the United States entered the war, there were three squadrons of Eagles, the 334th, 335th, and 336th Squadrons of the 4th Fighter Group of the US Eighth Air Force.

Spitfires took part in operations in the Middle East, North Africa, India, Burma, Australia, and the Soviet Union. The neutral governments of Portugal and Turkey were also equipped with Spitfires. When the war ended, the Spitfire was the only Allied airplane that had been in continuous production throughout the war—20,351 had rolled off the assembly lines.

The Spitfire on display in the World War II Aviation gallery in the National Air and Space Museum is a Mark VII, a high-altitude version of which only 140 were produced. It was transferred to the Smithsonian from the US Air Force in 1949.

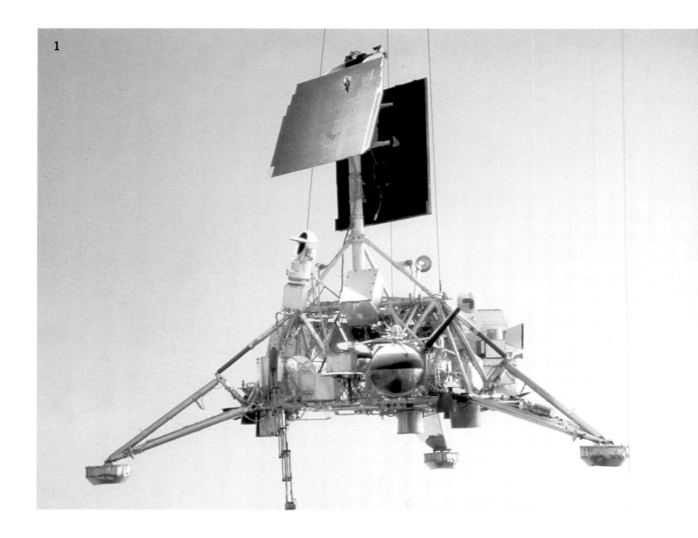

1

THE SURVEYOR PROBES were the first spacecraft to make soft landings on the Moon, and they sent 85,000 photographs back to Earth.

To find out information about the lunar surface, NASA sent three types of probes to the Moon: Ranger, Lunar Orbiter, and Surveyor. Unlike *Ranger* (see page 114), which crash-landed on the lunar surface, and Lunar Orbiter, which operated from lunar orbit, Surveyors were designed to soft land on and operate for extended periods of time directly on the Moon.

The objectives of the Surveyor program, as they were spelled out by NASA, were to accomplish a soft landing on the Moon, to provide basic data in support of the Apollo lunar landing program, and to perform scientific operations on the lunar surface. There were seven Surveyor launches, all of which used the Atlas-Centaur launch vehicle. *Surveyor 1, 3, 5, 6,* and *7* accomplished the goals set forth for the program. *Surveyor 2* and *4* were successfully launched but experienced problems en route to the Moon. As knowledge of the Moon and the capabilities of the spacecraft were obtained, changes were made to the Surveyor spacecraft and different landing sites were chosen.

Surveyor 1 was launched from Cape Kennedy, Florida, on May 30, 1966. All of the systems and procedures worked well. It landed on June 2, 1966, 16 km (10 mi.) north of Flamsteed Crater on the Moon's "Ocean of Storms." More than 10,000 high-quality, color photographs of the lunar surface were returned to Earth during the spacecraft's first day. *Surveyor 3* was similar to *Surveyor 1,* except that it had a scoop-and-claw device to test the

HEIGHT	3 M (10 FT.)
WIDTH	4.3 M (14 FT.)
WEIGHT	270 KG (596 LB.)

1. NASM's Surveyor lunar probe hanging on display in the Space Hall.
2. Astronaut Charles "Pete" Conrad Jr., *Apollo 12* commander, examines the *Surveyor 3;* the lunar module *Intrepid* is seen in the background.
3. Hughes Aircraft Co. technicians check the *Surveyor 1* spacecraft before its shipment to Cape Kennedy, Florida. **4.** A close-up view of the number 2 footpad of *Surveyor 3* taken during the Apollo 12 second moonwalk EVA on November 20, 1969.

lunar surface. It took the first color pictures of Earth from the Moon and made the first excavation on an extraterrestrial body. *Surveyor 5* was the first of the spacecraft to carry an alpha back-scattering instrument with which chemical analysis of the lunar material could be made.

Surveyor 6 made the first launch from the Moon when it hopped 2.5 m (8.2 ft.) so that its cameras and instruments could obtain data and images about the disturbance of the lunar soil caused by its initial landing. *Surveyor 7* was equipped with both the scoop-and-claw

device and the alpha back-scattering instrument. The arm was guided from Earth through a television camera.

Surveyor 3 had been on the Moon for two and a half years when the *Apollo 12* crew arrived in 1969. Astronauts Charles Conrad Jr. and Alan Bean removed its television camera, surface sampler, and some tubing and brought them back to Earth for analysis.

The Surveyor in the National Air and Space Museum is an engineering model constructed for thermal control studies. It was donated to the Smithsonian in 1968 and is on display in the Lunar Exploration Vehicles exhibit.

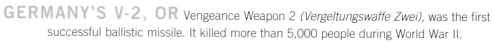

GERMANY'S V-2, OR Vengeance Weapon 2 *(Vergeltungswaffe Zwei),* was the first successful ballistic missile. It killed more than 5,000 people during World War II.

The Germans had been developing rocketry since the 1930s, aiming to create a long-range missile and exploring the use of rocket-powered aircraft. The liquid-propellant V-2 missile was first flown successfully from Peenemünde, Germany, in October 1942.

The V-2 was the largest and most complex missile in the German arsenal. It could

DIMENSIONS:

LENGTH	14 M (46 FT.)
WEIGHT	12,800 KG (28,000 LB.)
DIAMETER	1.65M (5 FT. 5 IN.)

1. The museum's V-2 Missile on display in the Space Race gallery. **2.** The launch of an A-4 (V-2) rocket from Peenemünde, Germany, circa 1943. **3.** A partially assembled German V-2 rocket on the assembly line at the underground plant at Nordhausen, Germany, shortly after the area was liberated by troops of the 1st US Army, April 10, 1945. **4.** Workers prepare a German V-2 rocket on the launch pad at Peenemünde, Germany.

send nearly 1,000 kg (1 ton) of explosives more than 270 km (170 mi.) in five minutes. The single rocket engine used a mixture of alcohol and liquid oxygen to provide thrust for about a minute. After engine shutoff, the missile traveled to its target on a ballistic trajectory—falling under the influence of gravity. The V-2 was guided during powered flight either by radio signals from the ground or by onboard gyroscopes and a device to measure the rocket's acceleration. It had control vanes in the rocket exhaust and air vanes on the fins.

Near the war's end, almost 600 V-2s were being produced monthly in caverns near Nordhausen by concentration camp prisoners under unbearably harsh working conditions. Thousands perished in the process.

From September 1944 to March 1945, about 2,900 V-2 missiles were fired against England, Belgium, and France from mobile launchers in Germany and its occupied territories. The V-2s were camouflaged to reduce their visibility to Allied bombers. More than 1,100 V-2s hit

southern England alone, causing an estimated 2,700 deaths and 6,500 injuries. Even more missiles were launched against the port city of Antwerp, Belgium. Because the Germans could not pinpoint targets with precision, anyone within the surrounding area could be hit.

As Allied armies liberated Europe in early 1945, American, French, British, and Soviet military intelligence teams raced to capture information, matériel, and personnel associated with German war technology. After surrendering to the US Army, Wernher von Braun, the head engineer for the V-2 project, and other V-2 experts revealed the capabilities of their rockets.

The V-2 missile in the National Air and Space Museum was reconstructed by the US Air Force using components from several missiles, and it was donated in 1954. It was painted to represent the first successful test missile fired from Peenemünde in October 1942. The markings made the missile easily visible for accurate assessment of its flight performance during testing.

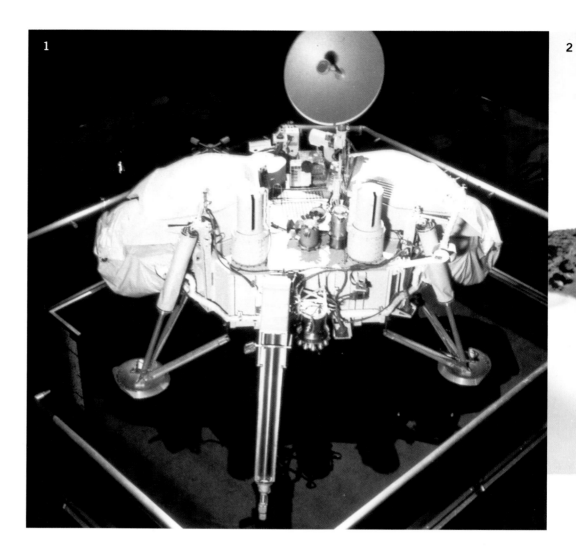

1

2

VIKING 1 AND 2 were the first spacecraft to conduct prolonged scientific studies on the surface of another planet.

Launched on August 20, 1975, *Viking 1* spent nearly a year cruising to Mars, then placed an orbiter in operation around the planet, and landed on July 20, 1976, on the Chryse Planitia (Golden Plains). *Viking 2* was launched on September 9, 1975, and landed on Mars on September 3, 1976.

The Viking project's primary mission ended on November 15, 1976, eleven days before Mars's superior conjunction (its passage behind the Sun), although the Viking spacecraft continued to operate for six more years. The last Viking transmission reached Earth on November 11, 1982.

The *Viking 1* and *2* orbiters studied Mars for six and four years, respectively, returning thousands of images of the planet. The two Viking landers eventually became the longest-surviving active laboratories on the surface of another world, far surpassing their original six-month design lifetime.

Just before entry into the atmosphere, the Viking lander was oriented so that the heat shield absorbed heat. At 6 km (4 mi.) above the surface, a parachute was deployed, and the aeroshell was then jettisoned. Finally, at 1.5 km (5,000 ft.), three radar-controlled retroengines were fired to keep the lander upright and further slow its descent. The landings occurred at a speed of 2 m/sec. (4.5 mph).

LENGTH	3 M (10 FT.)
HEIGHT	2 M (6 FT., 6 IN.)
WEIGHT	576 KG (1,270 LB.)

1. The museum's Viking Lander Spacecraft Test Article on display in Milestones of Flight gallery. 2. A view of the boulder-strewn landscape of Utopian Plain on Mars, taken by the *Viking 2* in November 1976. 3. First photograph ever taken on the surface of Mars taken by the *Viking 1* lander.

Scientific information on the Martian atmosphere was gathered even before the landers touched down. Starting at an altitude of 200 km (124 mi.), the instruments measured the composition and layering of the atmosphere. These experiments confirmed that carbon dioxide is presently the major component of the Martian atmosphere and that nitrogen may have been more abundant in the past.

Each lander was powered by two radioisotope thermoelectric generators, which convert heat to electricity from the radioactive decay of plutonium 238. The landers require 70 watts of power, less than what is needed by most lightbulbs.

Each Viking lander was equipped with two identical cameras that did not use film. Instead, a moveable mirror scanned a vertical segment of the Martian scene, and photodetectors recorded the amount of light reflected into the camera. A complete picture of the surface was made by completing a vertical scan and then rotating the camera slightly for the next scan.

Like a weather station on Earth, the Viking meteorology experiment measured the atmospheric pressure and temperature, and the wind speed and direction.

To determine the composition of the atmosphere and whether organic molecules exist in the surface samples, a gas chromatograph mass spectrometer was used. No organic molecules were detected at either landing site.

One of the most important scientific activities of this program involved determining whether there was life on Mars. Although the biology experiments discovered unexpected and enigmatic chemical activity in the Martian soil, they provided no clear evidence for the presence of living microorganisms in the soil.

During the radio science experiment, as Mars passed behind the Sun, the most accurate measurement of interplanetary distance was made. The 321-million-km (200-million-mi.) distance between Earth and Mars was measured to within 1.5 m (5 ft.), and the delay in radio signal due to the Sun's gravity was exactly that predicted by Albert Einstein's general theory of relativity—0.0002 seconds.

The Viking Mars lander in the National Air and Space Museum's Milestones of Flight gallery was used for performance tests and simulations before and during the actual mission. It was donated to the Smithsonian by NASA in 1979.

1

POWERED BY A Pratt & Whitney R-2800 engine, the Vought F4U Corsair had an immediate impact on the Pacific air war. Pilots used the Corsair's speed and firepower to engage the more maneuverable Japanese airplanes. Unprotected by armor or self-sealing fuel tanks, no Japanese fighter or bomber could withstand more than a few seconds of the concentrated volley from the Corsair's six .50-caliber machine guns.

Maj. Gregory "Pappy" Boyington assumed command of Marine Corsair squadron VMF-214, nicknamed the Black Sheep Squadron, on September 7, 1943. During less than five months of action, Boyington received credit for downing twenty-eight enemy aircraft. Enemy aircraft shot him down on January 3, 1944, but he survived the war in a Japanese prison camp.

In May and June 1944, Charles A. Lindbergh flew Corsair missions with Marine pilots at Green Island and Emirau. On September 3, 1944, Lindbergh demonstrated the F4U's bomb-hauling capacity by flying a Corsair from Marine Air Group 31 carrying three bombs each weighing 450 kg (1,000 lb.). He dropped this load on enemy positions at Wotje Atoll. On the September 8, Lindbergh dropped the first 900-kg (2,000-lb.) bomb during an attack on the atoll. For the finale five days later, he delivered a 900-kg (2,000-lb.) bomb and two 450-kg (1,000-lb.) bombs.

By the end of the war on September 2, 1945, the US Navy credited Corsair pilots with destroying 2,140 enemy aircraft in aerial combat. The navy and marines lost 189 F4Us in combat and 1,435 Corsairs in noncombat accidents. Beginning on February 13, 1942, US

DIMENSIONS: ———

WINGSPAN	12.5 M (41 FT.)
LENGTH	10.2 M (33 FT., 4 IN.)
HEIGHT	4.6 M (15 FT.)
WEIGHT	4,074 KG (8,982 LB.)

1. The museum's Vought F4U Corsair hanging on display at the Udvar-Hazy Center. The Curtiss P-40E, Naval Aircraft Factory N3F-3, and Lockheed SR-71 Blackbird can be seen in the background. 2. NASM's Vought F4U-1D Corsair *Sun Setter* displayed outside at the Paul E. Garber Facility following its restoration in October 1980. 3. A formation of four Vought F4U Corsairs in flight somewhere over the South Pacific in May 1943. 4. An official US Navy poster, "CORSAIRS CLIMB RIGHT INTO THE FIGHT!" by Jon Whitcomb depicts a Vought F4U Corsair flying over burning landscape.

Marine and Navy pilots flew 64,051 operational sorties, 54,470 from runways and 9,581 from carrier decks. During the war, the British Royal Navy accepted 2,012 Corsairs and the Royal New Zealand Air Force accepted 364. The demand was so great that the Goodyear Aircraft Corporation and the Brewster Aeronautical Corporation also produced the F4U.

Corsairs also flew from US Navy carrier decks and US Marine airfields during the Korean War. F4U pilots did not have many air-to-air encounters over Korea. Their primary mission was to support Allied ground units.

After World War II, civilian pilots adapted the speedy F4U to fly in competitive air races. They preferred modified versions of the F2G-1 and -2 originally built by Goodyear. Corsairs won the prestigious Thompson Trophy twice. In 1952, Vought manufactured 94 F4U-7s for the French navy, and these aircraft saw action over Indochina but this order marked the end of Corsair production. In production longer than any other US fighter to see service in World War II, Vought, Goodyear, and Brewster built a total of 12,582 F4Us.

The US Navy transferred an F4U-1D to the National Air and Space Museum in 1960. In 1980, NASM craftsmen restored the F4U-1D in the markings of a Corsair named *Sun Setter,* a fighter assigned to Marine Fighter Squadron VMF-114 when that unit served aboard the USS *Essex* in 1944.

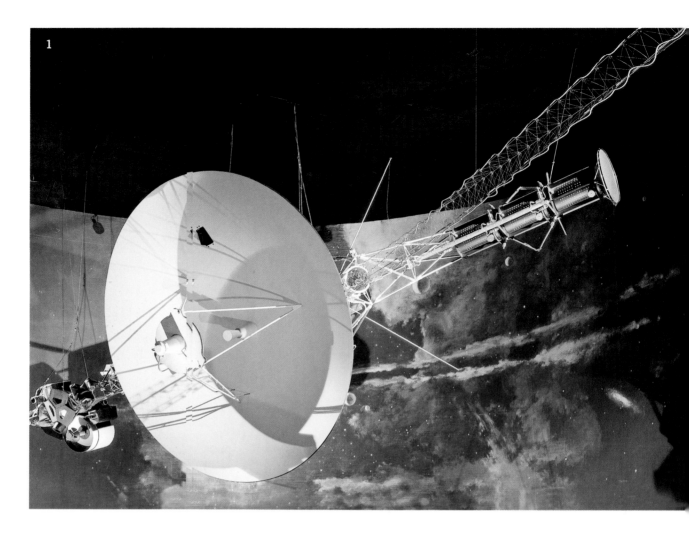

1

THE TWO VOYAGERS have taken over 100,000 images and returned information to Earth that has helped resolve some key questions and raise intriguing new ones about the evolution of the planets.

The *Voyager 1* and *2* spacecraft were originally supposed to investigate all of the outer planets in a "grand tour"—during a planetary alignment that will not occur again until 2157—but the mission was scaled back to only Jupiter and Saturn due to budgetary cutbacks.

Voyager 2 was launched on August 20, 1977, and placed on a slower flight path. It encountered Jupiter in July 1979 and Saturn in August 1981. Following a gravity assist by Saturn, *Voyager 2* continued on to explore Uranus in January 1986 and Neptune in August 1989. Pluto was not included in *Voyager 2*'s tour because its orbital position was out of range of the spacecraft's path.

Voyager 1 was launched on September 5, 1977, and reached Jupiter in March 1979 and then Saturn in November 1980. *Voyager 1*'s trajectory sent it within 4,104 km (2,550 mi.) of Saturn's moon Titan, the only other body in the solar system known to have a predominately nitrogen atmosphere like Earth's. This maneuver changed the spacecraft's flight path, sending it out of the solar system.

The two spacecraft have taken images of the outer planets, rings, and satellites, as well as millions of magnetic, chemical spectra, and radiation measurements. They discovered rings around Jupiter, volcanoes on Io, satellites in Saturn's rings, new moons around Uranus and Neptune, and geysers on Triton. For its last imaging sequence, *Voyager*

DIMENSIONS: ─────

DIAMETER	3.7 M (12 FT.)
HEIGHT	2.9 M (9 FT., 6 IN.)
LENGTH	6.5 M (21 FT.) SHORT BOOMS;
	17.5 M (57 FT.) DEPLOYED
	MAGNETOMETER BOOM
WEIGHT	861.5 KG (1,800 LB.)

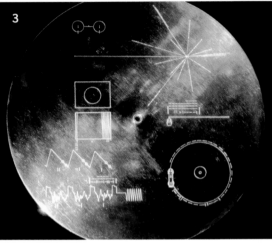

1. The museum's Voyager planetary probe spacecraft on exhibit in Exploring the Planets gallery.
2. A Voyager development model spacecraft is shown at NASA's Kennedy Space Center during transport and rocket shroud encapsulation tests. **3.** The Sounds of Earth Record Cover. The 12-inch gold-plated copper discs contain greetings in 60 languages, samples of music from different cultures and eras, and natural and man-made sounds from Earth. **4.** This montage of images of the Saturnian system taken by the *Voyager 1* was prepared for NASA by the Jet Propulsion Laboratory, Pasadena, California, in November 1980.

1 took a portrait of most of the solar system, showing Earth and six other planets as sparks in a dark sky.

Voyager 2 continues to travel away from our solar system, sending back information on interplanetary space. It continues to operate well, but its signals are growing fainter. If nothing happens to them, NASA should be able to remain in contact with the two Voyager space probes until about 2030. Both craft have enough hydrazine fuel to allow *Voyager 1* to keep traveling until 2040 and *Voyager 2* until 2034.

On November 5, 2003, the spacecraft reached 90 AU from the Sun, the equivalent of about 13.5 billion km (8.4 billion mi.). It is the only spacecraft to have made measurements in the solar wind from such a great distance from its source.

On the chance that someday they may be found by another civilization, *Voyager 1* and *2* both carry a copper phonograph record. Recorded on it are greetings in sixty languages, music from different cultures, and natural sounds such as those of wind, surf, and animals. The record also contains signals that can be converted into pictures.

Voyager is a Mariner-class spacecraft much like those that have explored Mercury, Venus, and Mars. Too little sunlight reaches the outer solar system for the Voyagers to use this solar power. Instead, they are equipped with radioisotope thermoelectric generators that convert heat (generated by radioactive decay of plutonium) into electricity.

Because the spacecraft travel at speeds in excess of 56,327 km/h (35,000 mph), and the light levels at Neptune can be 900 times fainter than those on Earth, the spacecraft angular rates were programmed to extremely small to prevent smearing.

Voyagers 1 and *2* have antennae 3.7 m (12 ft.) in diameter to transmit data over great distances. The Voyagers have transmitted data from as far as 9 billion km (5.7 billion mi.).

The Voyager in the National Air and Space Museum is a developmental test model that consists of facsimile and dummy parts. It was acquired by the Smithsonian in 1977 and placed on display in the Exploring the Planets gallery shortly thereafter.

AMERICA'S FIRST OPERATIONAL guided ballistic missile, the surface-to-surface Corporal (also designated SSM-A-17 and M2) was developed by the Jet Propulsion Lab (JPL) of the California Institute of Technology for the US Army and entered service in 1953. It remained in use until it was replaced by the longer-range, solid-fuel Sergeant missile in 1964.

DIMENSIONS: ═══════

LENGTH	14 M (46 FT.)
DIAMETER	.76 M (30 IN.)
SPAN	1.3 M (4.4 FT.)
WEIGHT	5,443 KG (12,000 LB.)

1. The museum's WAC Corporal missile on display in the Space Race gallery. **2.** Dr. Frank J. Molina (left, wearing hat) and two other unidentified men work on WAC Corporal missile at the Douglas Aircraft Company, Santa Monica, California. **3.** A WAC Corporal missile is prepared for launch.

The airframe was produced by the Firestone Tire and Rubber Company. The Ryan Aeronautical Corporation produced the liquid-propellant rocket motor to JPL specifications. Gilfillan Brothers, Inc., was responsible for the guidance system.

The Corporal missile had a range between 40 and 120 km (25 and 75 mi.) and could reach an altitude of 42 km (26 mi.) while on a ballistic trajectory. Its maximum speed was Mach 3.5.

The Corporal's 20,000-pound-thrust rocket engine used monoethylene and red fuming nitric acid and had a duration of 60 seconds. It could be equipped with either a conventional or atomic warhead and was suitable for tactical missions beyond the range of conventional artillery.

The Corporal was basically an artillery round launched vertically and partly guided by radar. The radar at the launch site established the target and a beam-rider guidance set the missile's ballistic trajectory. Minor course corrections were made by radar that actuated the gyro-controlled refractory graphite exhaust vanes.

The first successful Corporal was fired at White Sands proving grounds in New Mexico on May 22, 1947, and it was officially adopted as a tactical weapon in April 1954. The missile saw service overseas in US bases in West Germany, Italy, and the United Kingdom.

Its portable equipment included a lightweight launcher pedestal, erector, and guidance equipment trucks. It took six or seven hours to prepare a Corporal for launch. Because of the system's portability and, at that time, rapid deployment capability, it was of great use to the air force's tactical air command.

The Corporal missile donated to the National Air and Space Museum came through the US Navy's Disposal Division in 1967.

WITH THE WRIGHT 1903 Flyer, the Wright brothers created the first successful heavier-than-air flying machine and pioneered many of the tenets of modern aeronautical engineering.

In 1878, Wilbur and Orville Wright's father gave them a toy flying-helicopter model powered by strands of twisted rubber. They also had some experience with kites. In 1896, prompted by the fatal crash of glider pioneer Otto Lilienthal (see page 102), the Wright brothers began serious study of flight. After absorbing what materials the brothers had available locally, Wilbur wrote to the Smithsonian on May 30, 1899, requesting publications on aeronautics.

The Wrights built a 1.5-m (5-ft.) wingspan biplane kite in the summer of 1899. Their kite was used to test the viability of the control system that they planned to use in their first full-size glider. The Wrights accomplished effective lateral control by twisting, or warping, the tips of the wings in opposite directions via a series of lines attached to the outer edges of the wings that were manipulated by the pilot.

Encouraged by the success of their small wing-warping kite, the brothers built and flew two full-size piloted gliders in 1900 and 1901. Like the kite, these gliders were biplanes. For control of climb and descent, the gliders had forward-mounted horizontal stabilizers. Neither craft had a tail. The Wrights' hometown of Dayton, Ohio, was not suitable for flying gliders. An inquiry with the US Weather Bureau identified Kitty Hawk, North Carolina, with its sandy, wide-open spaces and strong, steady winds as an optimal test site. In September 1900, the Wright brothers made their first trip to the Outer Banks.

Although the control system worked well, the lift of the gliders was substantially less than the Wrights' earlier calculations had predicted. Questioning their data, Wilbur and

DIMENSIONS:	
WINGSPAN	12.3 M (40 FT., 4 IN.)
LENGTH	6.4 M (21 FT., 1 IN.)
HEIGHT	2.8 M (9 FT., 4 IN.)
WEIGHT	274 KG (605 LB.)

1. A frontal view of the Wright 1903 Flyer hanging on display in the Milestones of Flight gallery. **2.** "The first flight": 10:35 a.m., December 17, 1903, the Wright 1903 Flyer just after liftoff at Kill Devil Hill, Kitty Hawk, North Carolina. Orville Wright is at the controls as Wilbur Wright runs alongside. **3.** Wilbur (left) and Orville Wright (right) seated on porch steps of their family home in Dayton, Ohio, circa June 1909.

Orville decided to conduct extensive tests of wing shapes. They built a small wind tunnel in the fall of 1901 to gather accurate aerodynamic data for their next glider.

The Wrights' third glider, built in 1902 and based on the wind tunnel experiments, was a dramatic success. The lift problems were solved, and with a few refinements to the control system, they were able to make numerous extended controlled glides.

During the spring and summer of 1903 they built their first powered airplane. Essentially a larger and sturdier version of the 1902 glider, the only new component of the 1903 aircraft was the propulsion system. With the assistance of their bicycle-shop mechanic, Charles Taylor, the Wrights built a small, 12-hp gasoline engine. The truly innovative feature was the propellers. The brothers realized that propellers are rotating wings, producing a horizontal thrust force aerodynamically. The 1903 airplane was fitted with two propellers mounted behind the wings and connected to the engine, centrally located on the bottom wing, via a chain-and-sprocket transmission system.

By late 1903 the powered airplane was ready for trial. After winning the coin toss, Wilbur made an unsuccessful attempt on December 14, damaging the Flyer slightly. Repairs were completed for a second attempt on December 17. At 10:35 a.m. Orville lifted off for a 12-second flight, traveling 37 m (120 ft.). Three more flights were made that morning, the brothers alternating as pilot. The second and third were in the range of 61 m (200 ft.). With Wilbur at the controls, the fourth and last flight covered 260 m (852 ft.) in 59 seconds. There was no question that the Wrights had flown.

A gust of wind overturned and damaged the Wright Flyer, and it was never flown again. The Wrights built refined versions of the Flyer in 1904 and 1905, bringing the design to practicality.

The Wrights crated the 1903 Flyer and shipped it back to Dayton, where it remained in storage for more than a decade. Orville lent the Flyer to the Science Museum in London in 1928, where it stayed until it was formally donated to the Smithsonian on December 17, 1948, the forty-fifth anniversary of the flights. It has been on public display at the National Air and Space Museum ever since.

1

THE WRIGHT 1909 Military Flyer is the world's first military airplane.

In 1908 the US Army Signal Corps advertised for bids for a two-seat observation aircraft. The general requirements were that it had to be easy to assemble and disassemble so that an army wagon could transport it, it would have to carry two people with a combined weight of 160 kg (350 lb.), and it had to be able to fly 200 km (125 mi.) and reach a speed of at least 64 km/h (40 mph).

Furthermore, it had to be able to stay in the air for at least one hour without landing, and then land without causing any damage that prevented it from immediately starting another flight. It was required to ascend and land without requiring a specially prepared spot. And have the ability to land safely in case of a malfunction of the propulsion system. It had to be simple enough to operate after a reasonable amount of instruction. The purchase price was set at $25,000 with 10 percent added for each full mph of speed over the required 40 mph and 10 percent deducted for each full mile per hour under.

The Wright brothers constructed a two-place, wire-braced biplane with a 30- to 40-hp Wright vertical four-cylinder engine driving two wooden propellers, similar to the aircraft Wilbur had been demonstrating in Europe in 1908. This airplane made its first flight at Fort Myer, Virginia, on September 3, 1908. Orville set new duration records, including a 70-minute flight on September 11. He also made two flights with a passenger.

On September 17, however, tragedy occurred. At 5:14 p.m., Orville took off with

1. The Wright Type A Military Flyer (Signal Corps No.1) hanging on display in the Early Flight exhibition at the National Air and Space Museum. **2.** The Wright 1909 Military Flyer flies over the launch derrick during its flight trials at Fort Myer, Virginia, 1909. **3.** The Wright Type A Military Flyer, mounted for transport on top of a US Army wagon hitched to an automobile on September 1, 1908, passes by a hydrogen gas balloon probably belonging to the US Army Signal Corps No. 12. **4.** Orville and Wilbur Wright, Lt. Benjamin D. Foulois, and Lt. Frank P. Lahm refuel the Wright Type A Military Flyer in preparation for launch.

4

DIMENSIONS:

WINGSPAN	11.12 M (36 FT., 6 IN.)
LENGTH	8.82 M (28 FT., 11 IN.)
HEIGHT	2.46 M (8 FT., 1 IN.)
WEIGHT	333 KG (735 LB.)

Lt. Thomas O. Selfridge, the army's observer, as his passenger. The airplane had circled the field four and a half times when a propeller blade split. The aircraft, then at 46 m (150 ft.), safely glided to 23 m (75 ft.), when it then plunged to Earth. Orville was severely injured, including a broken hip, but Lieutenant Selfridge died from his injuries, and the aircraft was destroyed. Selfridge was the first person to perish in a powered airplane accident.

On June 3, 1909, the Wrights returned to Fort Myer with a new machine to complete the trials. The engine was the same as in the earlier aircraft, but the 1909 model had a smaller wing area and modifications to the rudder and the wire bracing. Lt. Frank P. Lahm and Lt. Benjamin D. Foulois, future army pilots, were the Wrights' passengers. On July 27, with Lahm, Orville made a record flight of 1 hour, 12 minutes, and 40 seconds, covering approximately 64 km (40 mi.), satisfying the army's endurance and passenger-carrying requirements. To establish the speed of

the airplane, Orville and Foulois made a 10-mile round-trip on July 30 at an average speed of 68 km/h (42.5 mph). For the 3 km/h (2 mph) over the required 40 mph, the Wrights earned an additional $5,000, making the final sale price of the airplane $30,000.

After taking possession of the Military Flyer, referred to as the Signal Corps No. 1 by the War Department, the Army conducted flight training at nearby College Park, Maryland, and at Fort Sam Houston in San Antonio, Texas, in 1910. Various modifications were made to the Military Flyer during this period. The most significant was the addition of wheels to the landing gear.

The War Department transferred the original Military Flyer to the Smithsonian on May 4, 1911. Apart from a few minor repairs, the airplane has not been restored, making it—out of three in the NASM collection—the one Wright airplane with the largest percentage of its original material and components.

CALBRAITH "CAL" PERRY Rodgers was the first to cross the United States by airplane in 1911 in his Wright EX biplane named *Vin Fiz*.

Publishing magnate William Randolph Hearst had announced a $50,000 prize for the first flight across the United States in thirty days or less. By September 1911, there were three competitors in the race—Rodgers, Robert Fowler, and James Ward. Fowler took off from San Francisco on September 11, but gave up by the end of the month. Ward took off from the East Coast on September 13, but withdrew a little more than a week later, not even making it out of New York State.

Rodgers secured financial backing from the Armour Company, which was introducing a new grape-flavored soft drink called Vin Fiz. Armour provided a special train, emblazoned with the Vin Fiz logo and cars for Rodgers's family and his support crew, and a hangar car, which was a rolling workshop filled with spare parts to repair and maintain the airplane. There was even an automobile on board to pick up Rodgers after forced landings away from the rail line.

Rodgers received five dollars for every mile he flew east of the Mississippi River and four dollars for every mile west of the river. Rodgers agreed to pay for the fuel, oil, spare parts, his mechanics, and the airplane itself, which the Wright Company agreed to build. Chief mechanic on the flight was Charles Taylor, who had worked for the Wright brothers since 1901.

The airplane was a Wright EX, a special design that was used for exhibition flying that was a slightly smaller version of the Wright Company's standard Model B flyer. Rodgers's aircraft carried the Vin Fiz logo on its wings and tail, and was quickly dubbed the Wright EX

DIMENSIONS:

WINGSPAN	9.6 M (31 FT., 6.5 IN.)
LENGTH	6.53 M (21 FT., 5 IN.)
HEIGHT	2.23 M (7 FT., 4 IN.)
WEIGHT	410 KG (903 LB.)

1. The Wright EX *Vin Fiz* hanging on public display in the Pioneers of Flight gallery. **2.** Cal Rodgers poses at the controls of his Wright EX *Vin Fiz* at the Pacific Ocean after having completed his transcontinental flight. **3.** A color poster depicting the route taken by Cal Rodgers on his 1911 transcontinental flight in the Wright EX *Vin Fiz*. **4.** Cal Rodgers takes off from Sheepshead Bay, New York, at the start of his 49-day coast-to-coast flight.

Vin Fiz. It was powered by a 35-hp Wright vertical four-cylinder engine and carried enough fuel for a maximum of three and a half hours of flying time.

Rodgers began his epic journey from Sheepshead Bay, New York, on September 17, 1911. The flight was punctuated by numerous stops, delays, and accidents. When the thirty-day time limit for the prize had expired, he had only reached Kansas City, Missouri. Undaunted, Rodgers continued on, whether he received the money or not. Upon leaving Kansas City, he flew south to Texas, and then across the southern US border toward Pasadena, California, the official termination point of the flight, while experiencing constant problems.

Trouble arose again on November 3, near Imperial Junction, California, less than 320 km (200 mi.) from the finish. At 1,200 m (4,000 ft.), an engine cylinder exploded, damaging one of the wings and driving steel shards into Rodgers's right arm. He struggled to regain control of the

Vin Fiz and, amazingly, managed to glide the 10 km (6 mi.) back to Imperial Junction and land safely. The engine and airplane were repaired a day later, and despite his painful injury, Rodgers departed for Pasadena, California, to a hero's welcome. The trip took forty-nine days, and he flew 6,914 km (4,321 mi.) for 82 hours total flying time at an average speed of 82.4 km/h (51.5 mph).

Although Pasadena was the official end of the coast-to-coast journey, Rodgers wanted to fly all the way to the Pacific shore. He took off from Pasadena for the short 37 km (23 mi.) trip to Long Beach on November 12. After a couple of mishaps and a crash that battered the EX and broke Rodgers's ankle, he reached the Pacific Ocean. To create a photo opportunity, *Vin Fiz* was rolled into the surf for the Pacific to lap over its wheels.

The *Vin Fiz* was donated to the Carnegie Institute in Pittsburg in 1917 by Rodgers's mother, and it was acquired by the Smithsonian in 1934.

PHOTOGRAPHY CREDITS

Key: l = left, r = right, c = center, t = top, b = bottom; DP = Dane A. Penland; EL = Eric F. Long; MA = Mark Avino; NASA = National Aeronautics and Space Administration; NASM = National Air and Space Museum; SI = Smithsonian Institution; USAF = United States Air Force; USN = United States Navy.

Front cover: EL, SI 2005-12594. **Back cover:** **tl** EL, SI 2005-24516; **tr** EL, SI 2002-19481; **c** DP, SI 2005-6025; **bl** DP, SI 79-763; **br** EL, SI 98-16042. **Front matter:** **i** DP, SI 2005-6458; **ii-iii** EL, SI 2002-19481; **iv-v** DP, SI 2005-6521; **vi-vii** EL, SI 2002-19565; **viii tl** EL, SI 2005-24516; **viii-ix** Dale Hrabak, SI 79-4639; **ix tr** DP/EL, SI 2005-22906; **x tl** DP, NASM 9A03171; **x-xi** DP, NASM 9A03248; **xi tr** EL, SI 97-16479; **xii** DP, SI 2004-31363; **xiv** DP, SI 2005-6275; **xvi** DP, SI 2004-19462. **Page 1 (2):** EL, SI 2001-128; **1 (3):** SI 74-8669; **2:** EL, SI 98-16042; **3 (2):** NASA-69-HC-789; **3 (3):** EL, SI 2005-6758; **4:** EL, SI 2005-11489; **5 (2):** NASA-AS11-40-5903; **5 (3):** NASA, SI 99-41160; **5 (4):** SI 92-616; **6:** EL, SI 2005-15152; **7 (2):** NASA, NASM 7B10257; **7 (3):** NASA, GPN-2000-001052; **8:** DP, SI 2005-423; **9 (2):** EL, SI 2001-125; **9 (3):** Photograph courtesy of the Imperial War Museum, London (IWM negative no. CL 3504); **9 (4):** SI 72-228; **10:** DP, SI 2005-24789; **11 (2):** Sándor Alex Àldott © Master of Photography, NASM 9A02865; **11 (3):** MA, SI 2004-11808; **11 (4):** Sándor Alex Àldott © Master of Photography, NASM 9A02866; **12:** EL, SI 2005-5389; **13 (2):** SI 75-5010; **13 (3):** EL, SI 2005-5391; **13 (4):** SI 90-8817; **14:** DP, SI 2004-17766; **15 (2):** © Bob Stidd, NASM 9A03142; **15 (3):** US Army News Features, SI 2006-624; **15 (4):** US Army News Features, NASM 9A001352; **16:** EL, SI 2005-24516; **17 (2):** EL, NASM 9A03155; **17 (3):** SI 97-17485; **17 (4):** USAF, NASM A-2013; **18:** EL, SI 97-16097; **19 (2):** Bell Helicopter Textron, SI 90-6683; **19 (3):** USAF, SI 87-7623; **20:** Dale Hrabak, SI 79-4639; **21 (2):** NASM 00180325; **21 (3):** SI 91-17304; **21 (4):** SI 2003-18720; **22:** DP, SI 80-2096; **23 (2):** SI 82-6974 ™ & © Boeing. Used under license; **23 (3):** SI 2003-33735; **24:** DP, SI 2005-6456; **25 (2):** SI 2005-3686; **25 (3):** SI 72-3667; **25 (4):** © American Airlines, NASM 9A00759; **26:** DP, SI 2005-5717; **27 (2):** SI 97-15276 ™ & © Boeing. Used under license; **27 (3):** NASM 7A07282 ™ & © Boeing. Used under license; **27 (4):** SI 88-14302 ™ & © Boeing. Used under license; **28:** DP, NASM 9A03248; **29 (2):** George E. Staley, SI 99-42697; **29 (3):** USAF, SI 2000-4554; **29 (4):** USAF, SI 2003-18753; **30:** DP, SI 80-4972; **31 (2):** SI 76-4861; **31 (3):** NASM A-4332 ™ & © Boeing. Used under license; **32:** DP, NASM 9A03171; **33 (2):** EL/MA, SI 2001-651; **33 (3):** USAF, NASM USAF-38361AC; **34:** © Breitling SA, NASM 9A02241; **35 (2):** Carolyn Russo, NASM 9A00390; **35 (3):** © Breitling SA, NASM 9A02242); **36:** EL, SI 2001-10002-12; **37 (2):** NASM A-51462; **37 (3):** Service Historique de l'Armée de l'Air, France (SHAA photo no. 83-5663); **37 (4):** DP, SI 2005-5190; **38:** EL, SI 2005-24467; **39 (2):** NASM 00169533; **39 (3):** B. H. Moody / Saudi Aramco / PADIA, SI 81-5078; **40:** DP, SI 2005-6275; **41 (2):** EL, SI 2003-20225-8; **41 (3):** Roland Leiser, NASM 9A01193; **41 (4):** EL, SI 2005-4060; **42:** EL, SI 2005-22899; **43 (2):** NASM A-38470-D; **43 (3):** NASM A-47151; **43 (4):** SI 89-5918; **44:** EL, NASM 9A02891; **45 (3):** SI 94-9641; **45 (3):** NASM A-38939-A; **46:** DP, SI 2004-18376; **47 (2):** Hans Groenhoff Photographic Collection, NASM HGC-638; **47 (3):** SI 2004-36599; **47 (4):** SI 79-3800; **48:** EL, SI 97-16073; **49 (2):** USAF, SI 2003-7; **49 (3):** NASM A-38882-A; **50:** DP, SI 2004-18886; **51 (2):** Federal Express, NASM 9A03174; **51 (3):** Federal Express, SI 76-3151; **52:** DP, SI 2004-18517; **53 (2):** Courtesy of Art Scholl Aviation, Inc., NASM 9A02164; **53 (3):** Courtesy of Art Scholl Aviation, Inc., NASM 9A02165; **53 (4):** Courtesy of Art Scholl Aviation, Inc., NASM 9A02162; **54:** EL, SI 2001-11467; **55 (2):** USN, SI A-53809; **55 (3):** NASA, SI 89-1881; **55 (4):** USN, SI 80-18590; **56:** EL, SI 2005-13984-11; **57 (2):** SI 72-6331; **57 (3):** SI 2005-3678; **57 (4):** USAF, SI 86-12548; **58:** DP, SI 80-4971; **59 (2):** Rudy Arnold Photo Collection, NASM XRA-0537; **59 (3):** USN, SI 74-8509; **59 (4):** USN, NASM 00040126; **60:** EL, SI 2001-1890; **61 (2):** SI 88-7415; **61 (3):** USAF, SI 83-390; **61 (4):** SI 83-364; **62:** DP, SI 80-4976; **63 (2):** NASM 9A02952; **63 (3):** NASM 7B14548; **63 (4):** SI 78-340; **64:** NASM 9A02841; **65 (2):** Lee Wells Collection, NASM A-43421; **65 (3):** Richard Hewitt Stewart/National Geographic Image Collection, SI 88-13351; **66:** MA, NASM 9A02849; **67 (2):** © Budd Davisson, SI 2001-1891; **67 (3):** Carolyn Russo, NASM 9A02844; **68:** Dale Hrabak, SI 83-14511; **69 (2):** USAF, NASM A-38634; **69 (3):** SI 87-3292; **69 (4):** SI 77-12503; **70:** EL, SI 2005-22896; **71 (2):** SI 80-9655; **71 (3):** SI Shell Companies Foundation, Inc., SI 89-5922; **71 (4):** SI 76-13308; **72:** MA, SI 97-15335; **73 (2):** Courtesy of Fairchild, SI 92-932; **73 (3):** SI 76-15516; **73 (4):** USAF, SI 98-15780; **74:** MA/EL, SI 97-15367; **75 (2):** SI 75-7267; **75 (3):** © American Airlines, SI 2001-5367; **76:** EL, SI 97-16479; **77 (2):** NASA-S65-30433; **77 (3):** NASA-S65-19528; **77 (4):** NASA-S65-19462; **78:** EL, SI 2005-17447; **79 (2):** NASM 00160386; **79 (3):** B. Anthony Stewart/National Geographic Image Collection, SI 89-21531; **80:** DP, SI 80-4973; **81 (2):** USN, SI 71-53-11; **81 (3):** NASM 9A00199; **81 (4):** NASM 9A00201; **82:** DP, NASM 9A03163; **83 (2):** USN, NASM 7A25957; **83 (3):** USN, SI 85-7306; **83 (4):** USN, SI 87-9666; **84:** DP, SI 2004-18854; **85 (2):** Hans Groenhoff Photo Collection, NASM HGD-035-14; **85 (3):** SI 75-5027; **85 (4):** NASM A-43509; **86:** DP, SI 2005-6566; **87 (2):** SI 91-7087; **87 (3):** Photograph courtesy of the Imperial War Museum, London (IWM negative no. HU 2408); **88:** EL, SI 97-16247; **89 (2):** NASA GPN-2000-001066; **89 (3):** NASA GPN-2000-001064; **90:** EL, SI 2005-4700; **91 (2):** SI 2002-14081; **91 (3):** SI 81-16962; **91 (4):** NASM A-46822-A; **92:** DP, SI 2004-18220; **93 (2):** SI 94-2155; **93 (3):** SI 91-13253; **93 (4):** SI 91-13252; **94:** NASM 9A03158; **95 (2):** EL, SI 2005-13630; **95 (3):** NASA, SI 2003-4850; **95 (4):** US Army, NASM 7B18821; **96:** Carolyn Russo, SI 2005-14079; **97 (2):** NASM A-47647-U; **97 (3):** USN, SI 72-3983; **97 (4):** EL, SI 2005-13983; **98:** DP, SI 2004-19461; **99 (2):** SI 2002-16636; **99 (3):** SI 2003-35050; **100:** DP, SI 2004-11706; **101 (2):** SI 77-912; **101 (3):** SI 84-13921; **101 (4):** Image provided courtesy of Bombardier Inc., NASM 00073319; **102:** EL, SI 2005-15502; **103 (2):** SI 2003-12097; **103 (3):** SI 2005-17747; **104:** DP, SI 80-2082; **105 (2):** SI 86-10744; **105 (3):** SI 80-11041; **105 (4):** SI 97-16118; **106:** DP, SI 2005-15710; **107 (2):** Rudy Arnold Photo Collection, SI 2002-17266; **107 (3):** Frank Griggs, NASM A-44235-D; **107 (4):** EL, SI 98-15012; **108:** EL, SI 2005-22902; **109 (2):** NASM A-48532-L; **109 (3):** Col. & Mrs. Charles A. Lindbergh, NASM A-45256-D; **109 (4):** SI 89-12246; **110:** DP, SI 2005-6025; **111 (3):** EL/MA, SI 2000-9346; **111 (3):** NASM 9A00308; **112:** EL, SI 2000-9403; **113 (2):** SI 85-7309; **113 (3):** NASM 9A02378; **113 (4):** Francis Gary Powers Papers, SI 93-16026; **114:** EL, SI 99-15232; **115 (2):** NASA GPN-2000-001131; **115 (3):** NASA-AS11-44-6642; **116:** EL, SI 2000-9371; **117 (2):** NASA Marshall Space Flight Center, MSFC-7022489; **117 (3):** EL, SI 99-15217-4; **117 (4):** NASA Marshall Space Flight Center, MSFC-7020219; **118:** EL, SI 2005-22898; **119 (2):** NASA, SI 84-6698; **119 (3):** NASM 7B04178; **119 (4):** MA, SI 97-15340; **120:** EL, SI 2005-24509; **121 (2):** NASA, NASM 7B11490; **121 (3):** NASA Jet Propulsion Laboratory, NASM 7B20523; **122:** DP, SI 2005-1520; **123 (2):** NASA GPN-2000-000461; **123 (3):** NASA GPN-2000-000484; **124:** EL, SI 97-15363; **125 (2):** USAF, SI 77-2694; **125 (3):** Ross Chapple, SI-2001-1899; **125 (4):** USAF, NASM A-42346; **126:** DP, SI 2004-18297; **127 (2):** NASM 9A00278; **127 (3):** SI 85-12340; **127 (4):** SI 90-16211 ™ & © Boeing. Used under license; **128:** EL, SI 97-16234; **129 (2):** NASA, NASM 9A02172; **129 (3):** NASA, SI 88-6821; **130:** Richard B. Farrar, SI 74-4295; **131 (2):** SI 73-1989; **131 (3):** SI 73-4119; **131 (4):** USAF, SI 90-4359; **132:** EL, SI 2005-22903; **133 (2):** SI 79-13829; **133 (3):** EL, SI 98-15875; **133 (4):** SI 76-13231; **134:** DP, SI 2004-18325; **135 (2):** USAF, SI 77-32; **135 (3):** SI 2003-36154; **135 (4):** EL, NASM 9A01656; **136:** EL, SI 2005-20395; **137 (2):** NASM 7A33871; **137 (3):** EL, NASM 9A01665; **137 (4):** USAF, NASM A-38634-C; **138:** DP, SI 2004-22934; **139 (2):** NASA, SI 2004-29402; **139 (3):** NASA, NASM 7B08913; **139 (4):** NASA, NASM 7B08920; **140:** DP, SI 2004-11948; **141 (2):** SI 98-16252; **141 (3):** SI 75-14972; **142:** DP, SI 2004-25989; **143 (2):** USAF, SI 80-5653; **143 (3):** Richard Rash, SI 97-1357; **143 (4):** Richard Rash, SI 2001-1889; **144:** DP, SI 2004-40580; **145 (2):** SI 93-1061; **145 (3):** USAF, SI 98-15407; **146:** DP, SI 79-833; **147 (2):** NASA, SI 71-2699; **147 (3):** SI 71-2699; **147 (4):** NASA Dryden Flight Research Center, EC61-0034; **148:** EL, SI 97-15873; **149 (2):** NASM A-34223; **149 (3):** SI 95-8736; **149 (4):** NASM A-47634-E; **150:** DP, NASM 9A02864; **151 (2):** NASM 7A36473; **151 (3):** © Northrop Grumman Corp., NASM 7A36466; **151 (4):** NASM 00078843; **152:** EL, SI 2005-20394; **153 (2):** NASA-72-HC-45; **153 (3):** © Northrop Grumman Corp., NASM 7B24172; **153 (4):** NASA, NASM 7B24147; **154:** MA, SI 2001-1877; **155 (2):** SI 83-16833; **155 (3):** Hans Groenhoff Photo Collection, SI 2004-51346; **155 (4):** SI 80-3908; **156:** NASM 9A00763; **157 (2):** NASM 9A00765; **157 (3):** EL, SI 2001-118; **157 (4):** SI 95-8289; **158:** EL, SI 2005-22904; **159 (2):** NASM 9A03252; **159 (3):** NASA, NASM 9A03254; **159 (4):** NASA, NASM 9A03256; **160:** DP/EL, SI 2005-22906; **161 (2):** SI 74-11710; **161 (3):** SI 2000-9714; **162:** DP, SI 2004-22945; **163 (2):** Eric Lundahl, NASM 7A38398; **163 (3):** Eric Lundahl, NASM 7A38397; **164:** MA, NASM 9A02843; **165 (2):** Doug Shane/Visions, NASM 9A02842; **165 (3):** SI 2001-7671; **165 (4):** MA, SI 87-16800-1; **166:** DP, SI 79-763; **167 (2):** NASM A-336; **167 (3):** SI 2003-28417; **167 (4):** EL, SI 2003-33193; **168:** EL, SI 2005-20396-1; **169 (2):** NASA S69-39309; **169 (3):** NASA Marshall Space Flight Center, MSFC-6862846; **169 (4):** NASA Marshall Space Flight Center, MSFC-6862832; **170:** DP, SI 2005-49; **171 (2):** Dorothy Cochrane, NASM 9A03258; **172:** NASA, NASM 9A01511; **173 (2):** EL, SI 2005-22900; **173 (3):** NASA SL3-115-1837; **173 (4):** NASA-73-HC-749; **174:** DP, SI 2004-40936; **175 (2):** NASA, SI 92-7437; **175 (3):** NASA, SI 2000-7592; **175 (4):** NASA-77-HC-111; **176:** DP, SI 2005-24437; **177 (2):** Courtesy of Scaled Composites, LLC, NASM 9A03260; **177 (3):** EL/MA, SI 2005-24957; **177 (4):** Courtesy of Scaled Composites, LLC, NASM 9A03263; **178:** MA, SI 86-12094; **179 (2):** MA/EL, SI 99-40421; **179 (3):** NASM 9A00823; **180:** EL, SI 97-15875; **181 (2):** SI 72-6833; **181 (3):** SI 79-14857; **181 (4):** USAF, SI 72-7660; **182:** SI 94-8282; **183 (2):** NASA, SI 98-15777; **183 (3):** NASA-66-H-342; **183 (4):** NASA-AS12-48-7110; **184:** EL, SI 97-16096; **185 (2):** SI 90-69; **185 (3):** Photograph courtesy of the Imperial War Museum, London (IWM negative no. OWIL 64336); **185 (4):** SI 77-14261; **186:** DP, SI 80-3070; **187 (2):** NASA, NASM 9A02856; **187 (3):** NASA, SI 2003-4993; **188:** DP, SI 2004-11951; **189 (2):** Dale Hrabak, SI 80-17164; **189 (3):** USN, NASM 7A09314; **189 (4):** NASM Poster Collection, SI 98-20683; **190:** EL, SI 2005-22901; **191 (2):** NASA-77-H-153; **191 (3):** NASA GPN-2000-001978; **191 (4):** NASA-80-HC-647; **192:** EL, SI 2005-17752; **193 (2):** SI 84-14740; **193 (3):** USAF, NASM USAF-32858AC; **194:** EL, SI 2002-19481; **195 (2):** SI 2003-3463; **195 (3):** SI 2002-16615; **196:** EL, SI 2005-20387; **197 (2):** SI 2003-29084; **197 (3):** Carl H. Claudy, SI 95-8433; **197 (4):** SI 93-13714; **198:** DP, SI 80-2081; **199 (2):** SI 77-9038; **199 (3):** NASM Poster Collection, SI 89-21352; **199 (4):** SI 2004-30408.

Page numbers in *italics* refer to illustrations.